FAIR BANANAS!

Gerardo Otero
Abril, 2011
North Vancouver

FAIR BANANAS!
Farmers, Workers, and Consumers
Strive to Change an Industry

HENRY J. FRUNDT

The University of Arizona Press   Tucson

*For Maria and my other grandchildren,
Michael, Nathaniel, Olivia, Luke, and Noah*

The University of Arizona Press
© 2009 The Arizona Board of Regents
All rights reserved

www.uapress.arizona.edu

Library of Congress Cataloging-in-Publication Data

Frundt, Henry J.
　Fair bananas! : farmers, workers, and consumers strive to change an industry / Henry J. Frundt.
　　p. cm.
　Includes bibliographical references and index.
　ISBN 978-0-8165-2720-5 (cloth : alk. paper) —
　ISBN 978-0-8165-2836-3 (pbk. : alk. paper)
　　1. Banana trade. I. Title. II. Title: Farmers, workers, and consumers strive to change an industry.
HD9259.B2F78　　2009
382'.414772—dc22　　　　2008048772

Publication of this book is made possible in part by the proceeds of a permanent endowment created with the assistance of a Challenge Grant from the National Endowment for the Humanities, a federal agency. The Ramapo College Foundation of New Jersey provided support for the paperback edition.

Manufactured in the United States of America on acid-free, archival-quality paper containing a minimum of 30% post-consumer waste and processed chlorine free.

14　13　12　11　10　09　　6　5　4　3　2　1

Contents

List of Illustrations   vii

Acknowledgments   ix

List of Acronyms   xi

1   Introduction: Competing Meanings of Fairness   1

2   Corporate Banana Structures   14

3   The Fair Trade Alternative   30

4   Actors for Banana Development   48

5   Going Bananas as a Social Movement   57

6   The Persistent Banana Environment   70

7   Conundrums of the Banana Trade   83

8   Resilience of Banana Unions   97

9   Peasants of the Caribbean and Fairer Trade   117

10 The Chiquita Accord and Labor Responses   137

11 The New Banana-Marketing Strategies   156

12 Fair Trade and Freedom of Association   173

13 Dole: Reluctant Fairness   189

14 A Proposal for Fairer Trade   197

15 Toward a Sustainable Banana Alliance   206

Notes   223

Bibliography   245

Index   267

Illustrations

Figures

1.1. Average wholesale banana prices   3
1.2. Nioka Abbott, banana farmer, St. Vincent   7
1.3. Union leader Ramón Barrantes at local headquarters   10
2.1. Del Monte banana worker, Guatemala   20
2.2. Chiquita and Del Monte stock performance, 1988–2005   23
3.1. Fair Trade logos   37
3.2. Percent stakeholder share in Fair Trade/conventional banana prices   42
6.1. Del Monte banana harvest, Guatemala   71
8.1. Negotiated worker housing, Sixaola, Costa Rica   103
8.2. Del Monte SITRIBI union committee under threat   111
8.3. Unemployed Chiquita contract workers, Guatemala   112
9.1. George DeFreitas, Fair Trade farmer, St. Vincent   131
9.2. Regina Joseph, Fair Trade farmer, Dominica   132
10.1. Chiquita workers discuss global labor rights framework accord   140
10.2. Unionized Chiquita workers, Sixaola, Costa Rica   141
13.1. FENACLE worker Muaro Romero in Luis Vernaza Hospital   190

Tables

1.1. Major National Banana Shipments by Weight and Value   4
2.1. TNC Positions in the Global Market   24
2.2. Top Five Grocery Retailers, USA   28
3.1. Fair Trade (FLO) Certification Standards   41
3.2. National Fair Trade Banana Sales   45
4.1. Key Fair-Banana Promoters   52
5.1. Definitions for a Social Alliance   58
6.1. Environmental Impacts of Monocultural Banana Production   72

7.1. Top Banana Importers, 2005   84
7.2. TNC Percentage Share of the European Banana Market   87
9.1. Caribbean Banana Exports by Weight and Value, 1995–2006   119
10.1. Banana Producer Race-to-the-Bottom Strategies   144
11.1. Banana Third-Party Certification Programs   158
11.2. Union Recommendations for Effective Codes   167
12.1. Fair Trade Production, Producers, Employees by Country, 2006   180
14.1. EUROBAN Proposals to Address Banana Trade Issues   203

Acknowledgments

I WANT TO EXPRESS my deep gratitude to the banana workers who contributed their views and insights to this study. I am likewise grateful to all those interviewed: union activists; small farmers; NGO representatives; government and corporate officials, including Ramón Barrantes, Irene Barrientos, Gilberth Bermúdez, Julio Coj, Rigoberto Dueñas, Warren Flores, Homero Fuentes, Reynoldo González, Sergio Guzman, Hernán Hermosilla, Carlos Mancilla, David Morales, Noe Ramirez, Aimee Shreck, Juan Tiney, Enrique Torres, Auria Vargas, Enrique Vázquez, German Zepeda, and many others. My special thanks go to all who helped review sections of the manuscript: Stephen Coats, USLEAP; Liz Parker, EUROBAN; Gelkha Buitrago, Veronica Perez, FLO; Roger Burbach, CENSA; Robert Armstrong, Steve Golin, Karen Judd, and my marvelous wife, Bette. I also received assistance from my students and colleagues at Ramapo College and the American Federation of Teachers, notably Yolanda Prieto, Charles Carreras, Martha Ecker, Ray Fallon, Ronald Kase, Ellen Ross, Trent Schroyer, as well as my friends in Central America, Mario Aníbal González, Maria Elena Anzueto, Raul Anzueto, Miguel Aczoc, Mariel Aguilar, Ana Arellano, Rolando Cabrera, Arlena Cifuentes, Veronica González, Dale Johnson, Frank LaRue, Carrie McCracken, Robert Perillo, Victor Quezada, Richard Spohn, Maria Trejos, students and colleagues at the Universidad Rafael Landivar, Guatemala, and many more. My thorough appreciation goes to Allyson Carter, Nancy Arora, Kenneth Plax, Alice Landwehr and the editors and anonymous reviewers at the University of Arizona Press who kept insisting on improvements. Trinidad photographer Abagail Hadeed graciously made available several of her lovely pictures. I am indebted to the Ramapo College Foundation for its generous support for this project. I also express my appreciation to the University Seminars at Columbia University, New York, for their help in publication.

The ideas presented here have considerably benefited from discussions in the University Seminar on Globalization, Labor, and Popular Struggles, co-chaired by David Bensman and myself. To the reader, I humbly beg your tolerance for my limited ability to present the intricate justice of the banana struggle.

Acronyms

ACILS	American Center for International Labor Solidarity (The Solidarity Center)
ACP	African, Caribbean, and Pacific
AECO	Asociación Ecologista Costarricense–Amigos de la Tierra (AECO-AT) (Costa Rican Ecologist Association–Friends of the Earth)
AIFLD	American Institute for Free Labor Development
AK–NAFLU–KMU	Amado Kadena–National Federation of Labor Unions–Kilusang Mayu Uno (Philippines)
APOQ	Asocición de Pequeños Productores de Orgánicos de Querecotillo (Asociation of Small Organic Producers of Querecotillo)
APPBOSA	Asociación de Pequeños Productores de Bananas Orgánicos de Saman y Anexos (Association of Small Organic Banana Producers of Saman and related)
APROVOPCHIRA	Asociación de Productores de Banano del Valle de Chira (Banana Producer Association of the Chira Valley)
APTTA	Asociación de Pequeños Productores de Talamanca (Association of Small Producers of Talamanca, Costa Rica)
ASDA	(Wal-Mart-owned U.K. discount food chain founded by 1965 merger of Asquith and Associated Dairy)
ASEPROLA	Asociación Servicios de Promoción Laboral (Labor Promotion Service Association, Costa Rica)
ASOPROBAN	Cooperativa de Parceleros y Pequeños Productores (Association of Smallholders and Producers, Colombia)
ASOTRAMA	Asociación Pro Defensa de los Trabajadores Agrícolas y del Medio Ambiente (Association for

	the Protection of Agricultural Workers and the Environment, Costa Rica)
ATC	Asociación de Trabajadores del Campo (Association of Rural Workers, Nicaragua)
AUC	Audodefensas Unidas de Colombia (United Self-Defense Forces of Colombia)
BANASAN	Group of five Colombian firms, Tropic S.A., Banapalm, Banex, Frutesa, and Agramayor
BANDECO	Banana Development Company (Del Monte Subsidiary, Costa Rica)
BANDEGUA	Compañia de Desarrollo Bananero de Guatemala Limitada (Del Monte Subsidiary, Guatemala)
BIOCOSTA SAC	Peruvian Fair Trade Exporter
BVQT	Bureau Veritas Quality International, now Bureau Veritas Certification
CAFOD	Catholic Agency for Overseas Development
CARICOM	Caribbean Community and Common Market
CARIFORUM	Caribbean Forum of African, Caribbean and Pacific States
CEPIBO	Central Piurana de Asociaciones de Pequeños Productores de Banano Orgánico (Central of Peruvian Associations of Small Organic Banana Producers)
CERAI	Central de Estudios Rurales de Agricultura Internacional (Center of Rural Studies of International Agriculture)
CGT	Conféderation Génèrale de Travail (General Confederation of Labor, France)
CLAC	Coordinadora Latinoamericana y del Caribe de Pequeños Productores de Comercio Justo (Latin American and Caribbean Network of Small Fair Trade Producers)
COAG	Coordinadora de Organizaciones de Agricultures y Ganaderos (Coordination of Organizations of Farmers and Ranchers, Spain)
COBIGUA	Corporación Bananera Guatemalteca Independiente (Guatemalan Independent Banana Corporation [Independent Producers for Chiquita])
COBAL	Compañia Bananera Atlántica (Atlantic Banana Company, Chiquita Subsidiary, Costa Rica)

COPDEBAN	Corporación Peruana de Desarrollo Bananero (Peruvian Corporation for Banana Development, or Dole Peru)
COLSIBA	Coordinadora Latinoamericana de Sindicatos Bananeros (Regional Coordination of Latin American Banana Workers' Unions)
COSIBA-CR	Coordinadora Sindicatos Bananeros de Costa Rica (Coordination of Costa Rican Banana Unions)
COOBANA	Cooperativo de Bananeros de Bocas del Toro (Cooperative of Bananeros of Bocas del Toro)
COOPSEMUPAR	Cooperativo de Servicios Multiples de Puerto Armuelles (Puerto Armuelles Fruit Company Multiple Services Cooperative)
CORBANA	Corporación Bananera Nacional (National Banana Export Association, Costa Rica)
COSIBAH	Coordinadora Sindicatos Bananeros y Agroindustrias de Honduras (Coordination of Banana and Agroindustrial Workers Unions of Honduras)
CPASM or Coopetrabasur	CPASM de Trabajadores Bananeros del Sur (Cooperative of Banana Workers of the South, Costa Rica)
CTM Altromercato	Italy's largest alternative trading organization
CUSG	Confederación Unidad de Sindicatos de Guatemala (United Confederation of Guatemalan Unions)
CUT	Central Unitaria de Trabajadores de Colombia (Unitary Workers Confederation, Colombia)
DBCP	Dibromochloropropane, a chemical treatment for nemátodes
EFTA	European Fair Trade Association
ELN	Ejército de Liberación Nacional (National Liberation Army, Colombia)
EPA	Environmental Protection Agency (U.S.)
EPAs	economic-partnership agreements between governments (or between the EU and ACP governments)
ETI	Ethical Trade Initiative
EUROBAN	European Banana Action Network
FAO	Food and Agricultural Organization of the United Nations

FAO/ITC/CTA	Food and Agricultural Organization of the United Nations/International Trade Commission/Commodity Trading Advisors
FAOSTAT	FAO Statistics
FARC	Fuerzas Armadas Revolucionarias de Colombia (Columbian Revolutionary Armed Forces)
FDP	Fresh Del Monte Produce
FENACLE	Federación Nacional de Trabajadores Agroindustriales de Campesinos e Indígenas Libres del Ecuador (National Federation of Small Farmers and Indigenous Community Organizations of Ecuador)
FESTRAS	Federación Sindical de Trabajadores de la Alimentación, Agroindustria y Similares (Guatemalan Food Workers Federation)
FETRABACH	Federación de Trabajadores Bananeros de Chinandega (Federation of Banana Workers of Chinandega, Nicaragua)
FINE	Joint alternative trade advocacy organization for Fair Trade, IFAT, NEWS! and EFTA, based in Brussels, Belgium
FLACSO/CEDAL/FES	Facultad Latinoamericana de Ciencias Sociales/Centro de Estudios Democráticos de América Latina/Friedrich Ebert Stiftung Foundation
FLO	Fairtrade Labelling Organizations International
FLO-CERT	Fairtrade Labelling Organizations–Certification
FNV	Federatie Nederlandse Vakbeweging (Federated Netherlands Labor Movement), the country's largest federation of trade unions
FOA	freedom of association
FOB	free on board
FT	Fair Trade
FTF	Fair Trade Foundation (U.K.)
GATT	General Agreement on Tariffs and Trade
GAWU	Ghana Agricultural Workers Union
GSP	Generalized System of Preferences
IBCII	International Banana Conference II
ICM	Integrated Crop Management
IFAT	International Federation of Alternative Trade

IFOAM	International Federation of Organic Agricultural Movements
IISD	International Institute for Sustainable Development
ILO	International Labor Organization
INABAB	International Network for the Improvement of Banana and Plantain, Montpellier, France
ISCOD	Instituto Sindical de Cooperación al Desarrollo (Union Institute for Cooperation and Development, Spain)
ISEAL	International Social and Environmental Accreditation and Labelling Alliance
ISO	International Organization for Standardization
ITGLWF	International Textile, Garment and Leather Workers Federation
IUF	International Union of Food, Agricultural, Hotel, Restaurant, Catering, Tobacco and Allied Workers Associations
LAWG	Latin American Working Group
MSN	Maquila Solidarity Network
NAALC	North American Agreement on Labor Cooperation (NAFTA side agreement between the United States, Mexico, and Canada)
NAFTA	North American Free Trade Agreement
NEWS!	Network of European Worldshops
NGO	nongovernmental organization
NISGUA	Network in Solidarity with the People of Guatemala
NREA	National Economic Research Associates
NSM	New Social Movement
RA	Rainforest Alliance
REPEBAN	Red de Pequeños Productores de Banano de Comercio Justo (Peruvian Network of Small FT Banana Producers, including Appbosa)
RONGEAD	a network of European NGOs on agriculture, food, trade, environment, and development
SAI	Social Accountability International
SAN	Sustainable Agricultural Network
SERAGRO-SAC	Dole-created labor contracting company in Peru
SERRV	Sales Exchange for Refugee Rehabilitation and Vocation

SGS	Société Gènérale de Surveillance
SINTAGRO	Sindicato de Trabajadores Agrícolas (Union of Agricultural Workers)
SINTRABANANO	Sindicato de Trabajadores Bananeros (Union of Banana Workers, Colombia)
SINTRASPLENDOR	Sindicato de Trabajadores de Splendor (Union of Splendor Flower Workers, Colombia)
SITAGAH	Sindicato de Trabajadores de Plantaciones Agrícolas y Ganaderos de Heredia (Union of Livestock and Agricultural Plantation Workers of Heredia, Costa Rica)
SITAG	Sindicatos de Trabajadores Agrícolas de Perú (Unions of Agricultural Workers of Peru)
SITRAAMERIBI	Sindicato de Trabajadores de Ameribi (Union of Workers of the Ameribi Plantation, Honduras)
SITRABI	Sindicato de Trabajadores Bananeros de Izabal (Union of Banana Workers of Izabal, Guatemala)
SITRABI	Sindicatos de Trabajadores Independientes de Panama (Unions of Independent Workers, Bocas del Toro, Panama)
SITRACHILCO	Sindicato de Trabajadores de Chilco (Union of Workers of Chilco, Puerto Armuelles, Panama)
SITRAIBANA	Sindicato de Trabajadores de la Industria Bananera (Union of Workers of the Banana Industry, Changuinola, Panamá)
SITRAINAGRO	Sindicato Nacional de Trabajadores de la Industria Agropecuaria (Union of Workers in Industrial Agriculture, Colombia)
SITRAP	Sindicato de Trabajadores de Plantaciones Agrícolas (Union of Agricultural Plantation Workers, Siquirres, Costa Rica)
SITRATERCO	Sindicato de Trabajadores de la Tela Railroad Company (Union of Workers of the Tela Railroad Company, Honduras)
SLBC	St. Lucia Banana Corporation
SOPISCO	newsletter covering banana shipping business based in Ecuador
STITCH	Support Team International for Textileras
TNC	transnational corporation
TQF	Tropical Quality Fruit Company, St. Lucia

UBESA	Unión de Bananeros Ecuatorianos, SA (Union of Ecuadorian Banana Producers [Dole's Ecuadorian Division])
UCIRI	Unión de Comunidades Indígenas de la Región del Istmo (Union of Indigenous Communities of the Isthmus Region, Oaxaca, Mexico)
UFCW	United Food and Commercial Workers Union, U.S.
UFW	United Farm Workers
UNCTAD	United Nations Conference on Trade and Development
UNITE-HERE	Union of Needletrades, Industrial and Textile Employees–Hotel Employees and Restaurant Employees International Union
UNSITRAGUA	Unión Sindical de Trabajadores de Guatemala (Union of Independent Workers, Guatemala)
UROCAL	Unión Regional de Organizaciones Campesinas del Litoral (Regional Union of Peasant Organizations of Litoral, Ecuador)
USAS	United Students Against Sweatshops
USLEAP	U.S. Labor Education in the Americas Project
USTR	U.S. Trade Representative
VREL	Volta River Estates, Ltd., Ghana
WIBDECO	Windward Island Banana Development and Export Company
WINFA	Windward Islands Farmers Association
WTO	World Trade Organization

FAIR BANANAS!

CHAPTER 1

Introduction
Competing Meanings of Fairness

WHEN IT COMES to fruit, we choose bananas the most. For a century this curved delight has graced northern breakfast tables. We buy bananas far more often than apples, oranges, or grapes, making them half of all the fruit purchased around the globe. Each of us ingests about twenty-eight pounds a year with satisfaction, reaping potassium, amino acids, fiber, and vitamin C. Yet until recently, we have paid scant attention to the 10 million people involved in growing and packing these bananas.

In the United States, we are awakening to the millions of farmers and workers involved, inspired in part by the European quest for "fair trade"—the purchase of sustainably produced goods on terms that ensure an adequate livelihood to producers. As we encounter the opportunity to purchase fair-trade products, including bananas, thousands of U.S. consumers are now considering the impact of their food choices on underdeveloped societies.

Individuals and organizations that stress equitable and sustainable development find this rising consciousness laudable. If customers are increasingly informed about who grows or makes what they buy, if people shop knowledgeably for goods that assure producers an adequate livelihood, world conflict may lessen. Nevertheless, even as committed consumers and necessary development organizations, we may oversimplify the complex reality of global production and trade.

Banana growing and marketing are challenging examples of such complexity. For much of the past century, the fruit sold in the United States came from corporate-controlled plantations in Latin America, where workers labored long and arduous hours, enduring environmental risks in hot climates for little pay. By contrast, most bananas sold in England, France, Spain, Portugal, and Greece were grown by peasants in the Caribbean, the Mediterranean, or the Canary Islands who also labored strenuously but for somewhat better pay than workers in the Americas. These European nations had created an early fair-trade system of tariffs and quotas for bananas, one that many advocates would like to revive and preserve.

Regardless of the export regime, however, banana production and trade remained a difficult business. In clearing swaths of land, producers often displaced small farmers and destroyed forests and wildlife. Wherever they handled fruit, corporate shippers demanded more efficient rules for planting, harvesting, and packing, and introduced chemicals that wrought havoc on habitat, workers, and communities. To assure seedlings irrigation and drainage, shippers redesigned waterways in a manner that increased soil erosion. Bananas became an extensive monocultural crop that supplanted diverse plant species, becoming more susceptible to insects, fungi, and competing weeds that workers and small producers were expected to broadside with pesticides. Corporate overseers imposed more intense work requirements, resulting in lower compensation for those who worked the land and packed the produce. Despite company innovations, workers still carried backbreaking loads and endured toxic risks.

As shipments increased during the past century, Latin workers fought these conditions tenaciously and ultimately won collective labor agreements that brought improvements. However, as companies and governments emphasized trade competition in the century's latter decades, European nations gradually obliterated guarantees for small banana producers. Corporations sought cheaper production. Often this meant shifting cultivation from an area, bypassing environmental regulations, and undercutting smallholder and labor protections.[1] Although banana producers increased the global output of export bananas by over 30 percent between 1990 and 2005, and real average wholesale prices rose above 35¢ a pound through 1996, prices subsequently dropped notably (see fig. 1.1).[2]

Thus, despite Europe, North America and Japan's—the North's—love for bananas and consequent increase in imports, lower prices in the twenty-first century have exacerbated the difficulties faced by workers and small producers in the South. In table 1.1 I compare exports of the top ten banana-shipping countries from 1995 to 1999 with those from 2000 to 2004. Note declines from Honduras and Panama, where unions have been notably strong. Small-producer exports from the Caribbean's

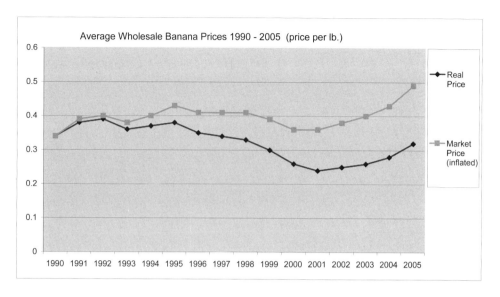

FIGURE 1.1. Average wholesale banana prices, 1990–2005 (real prices calculated using U.S. inflation rates).

Windward Islands also dropped dramatically (see table 9.1). Yet as table 1.1 also shows, overall banana output generally expanded. As a consequence of the increase in supply, prices, then wages, have fallen in what observers call a "race to the bottom" as competition among banana shippers has intensified. Iain Farquhar, of Banana Link, a British group dedicated to supporting banana workers, compared this race to the bottom to a bungee jump in which no one remembers to secure the bungee (Banana Link 2005).

Companies began delivering bananas at bargain rates, but as customers learned about paltry wages and practices, they questioned how much fruit cultivation benefited local producers. As arduous work conditions and detrimental environmental practices became more widely publicized, consumers expressed amazement that smallholders and workers received such a tiny portion of the price they paid in the supermarket, often less than 2 percent (see fig. 3.2). Consumers rightfully demanded fairer bananas. Yet the complexities of the current banana-production system render true fairness difficult to achieve.

TABLE 1.1. Major National Banana Shipments by Weight and Value 1995–2006 (metric tons and 000 dollars)

Country		1995–99 avg.	2000–04 avg.	2005	Percentage change, 2000–04	2006	Percentage change from 2005
Brazil	mt	46,455	165,495	212,180	0.28	194,300	−0.08
	$	8,517	23,814	33,027	0.39	38,460	0.16
Cameroon	mt	165,237	267,858	265,460	−0.01	256,630	−0.03
	$	53,490	57,494	68,236	0.19	64,321	−0.06
Colombia	mt	1,505,834	1,453,018	1,621,750	0.12	1,567,900	−0.03
	$	468,549	399,113	464,959	0.16	468,549	0.04
Costa Rica	mt	2,133,742	1,994,231	1,775,520	−0.11	2,183,510	0.23
	$	642,865	528,644	483,492	−0.09	634,144	0.31
Ecuador	mt	3,963,026	4,273,965	4,764,193	0.11	4,908,564	0.03
	$	1,019,718	926,320	1,068,659	0.15	1,184,355	0.11
Guatemala	mt	664,780	930,035	1,129,477	0.21	1,055,497	−0.07
	$	154,325	201,922	238,100	0.18	216,808	−0.09
Honduras	mt	432,795	453,965	545,527	0.20	515,224	−0.06
	$	106,396	136,450	134,698	−0.01	130,835	−0.03
Ivory Coast	mt	204,532	249,896	234,267	−0.06	286,301	0.22
	$	76,480	80,394	94,351	0.17	115,346	0.22
Panama	mt	597,191	420,510	352,480	−0.16	431,141	0.22
	$	175,047	118,662	96,517	−0.19	109,073	0.13
Philippines	mt	1,215,625	1,808,188	2,024,321	0.12	2,311,540	0.14
	$	226,862	311,467	430,000	0.38	405,444	−0.06
World	mt	14,001,155	14,872,695	16,205,664	0.09	16,789,032	0.04
	$	4,835,585	4,476,663	5,653,926	0.26	5,799,147	0.03

Source: Computed from FAOSTAT.FAO.org (detailed trade statistics) 2008.

Fair Trade Definitions

When consumers buy fruit, they seek "fair bananas," that is, the assurance that the environment remains healthy and producers receive an adequate income for their role in the overall system of production and distribution. However, the route to fairly traded bananas can be a matter of debate. For some, fair trade can happen only when the free-market capitalist system is either eliminated, or totally overhauled to reverse the unequal terms of trade it generates between northern consumers and southern producers.[3] The majority of customers, however, accepts some form of capitalist trading. Many activists in Europe would restore the tariff/quota trading system that protected small banana farmers in former African, Caribbean, and Pacific (ACP) colonies that sold to designated markets in the North. Following up on this desire, certain corporate shippers, and small producers dependent on them, define fair trade as their historical claim to licenses and guarantees codified by European governments. For many banana workers in Latin America, however, fairness simply signifies the assurance of just wages and working conditions, best achieved by respecting their right to free-functioning unions. Newer claims on the meaning of fair trade have emerged from nationally based independent producers who believe that if they are given greater access to global sales, they can offer farmers and workers better pay. Adding their voices to the debate, environmentalists point out that fairness means respecting the long-term health and well-being of land and communities.[4]

A Hypothesis

While these meanings of fair bananas span a diverse range, I hypothesize in this book that the groups articulating them have the potential for coalescing around a set of principles. I envision a convergence among small farmers, banana workers, union leaders, and activist NGOs. Personal examples begin to illustrate this convergence: George DeFreitas, Nioka Abbott, Anton Bowman, and Juan Quenteña offer smallholder perspectives; workers like Jiménez Guerra and Auria Vargas offer union views, as do leaders like Ramón Barrantes. Retail promoters like Jonathan Rosenthal, past director of Oké Bananas, present yet another perspective.

Smallholders

George DeFreitas is a small producer in the Windward Islands, where growing bananas costs more than twice what it does in Ecuador:

> St. Vincent is completely dependent on bananas. Whereas other crops might only be harvested once or twice a year, bananas give people a weekly income.... There are still some younger people on the farms but lots have left. Farmers paint such a gloomy picture of what it's like that the youth don't want to get involved. We depend heavily on being able to sell at a good price to a good market. If there was a growth in the market for bananas then more people would get involved in production again. (Oxfam 2004)

Along with George, Nioka Abbott has been in the St. Vincent banana business. Since 1989, Nioka has cut and packed fruit from her two leased acres every one or two weeks. "Bananas are better than any other crops for regular harvesting. You get an income all through the year," she says. "That's why the banana is so popular as a cash crop" (Oxfam 2004). George, Nioka, and their coproducers urge support for a fair-trade system that offers a viable livelihood (fig. 1.2).

Anton Bowman is an older, wiry man with a merry disposition who runs a banana farm on the Windward Island of St. Vincent and leads a farmers' association there. Anton had been a trade unionist before he changed occupations and became self-employed. Now, he hires other workers to grow bananas. He actively promotes the interests of banana producers through the Windward Islands Farmers Association (WINFA), to which many other smallholder producers, such as George DeFreitas and Nioka Abbott, belong. Anton advocates fairer trade and the social and environmental benefits it brings; but he also seeks broader connections to confront the myriad difficulties faced by island holders and the banana sector overall (Bowman 2005).

Juan Quenteña is a small banana producer in Ecuador who sells through a friendly middleman to the Favorita company, an affiliate of the giant transnational firm Chiquita. Juan owns ten hectares (about twenty-two acres) devoted to bananas. With help, Juan harvests about 450 boxes a week. Although other independents may pay more, he prefers Favorita because it is a consistent buyer, although it charges for fertilizer

FIGURE 1.2. Nioka Abbott, banana farmer, St. Vincent. (Photo by Abigail Hadeed, Trinidad)

and aerial spraying. In 2001, he received $1.16 below the government's official minimum box price and $4 below the average world market price. Fertilizer cost $216/ha; spraying against leaf disease, another $390/ha (Hellin and Higman 2001). Juan's additional infrastructure expenses varied between $1 and $2 a box. Although he finds bananas more dependable than the cacao he had been growing before, Juan still needs at least $2.50 a box to cover expenses and give him an adequate income. A fairly traded banana would allow him to meet costs, pay a livable wage, and minimize environmental damage.

Workers

Banana field hands also have a stake in a "fair banana." Jiménez Guerra, a well-built, perceptive indigenous worker, has labored in Chiquita's Costa Rican banana fields since 1995. He often works twelve hours a day cutting off leaves infected with a black fungus known as sigatoka. "Our union contract is not great," says Guerra. "I receive 3600 colones (about $9) for a day's work. Even when the work is very hard, they won't pay more. This kind of wage is not enough for raising a family. Benefits do not cover all health costs. Environmentally, things have improved, yet about once a month, the company sprays the fields without giving us time to leave" (Guerra 2004).

For Jiménez, fairer trade would bring workers an adequate wage and better conditions. Nine dollars a day for hard work and the risk of a monthly toxic spray is not a friendly situation. But then he compares this to the conditions faced by fellow and sister workers at independent *fincas* (plantations) nearby that *sell* to Chiquita, who are paid only 1,100 colones ($2.80!) for eight hours of work—less than a third of what Jiménez earns. This latter group filed an official complaint about minimum-wage violations. "They pay us 14 cents for each hundred-pound stem of bananas that we cut, carry to the cable, and transport to the packing shed," states Enrique, who walks in each day from Panama. Enrique and the others describe how they are showered with chemical sprays at least once a week! (interview with Workers of Sixaola 2004).

Packing plant workers also have to contend with toxic conditions. Auria Vargas Castañeda, the Chiquita Sixaola union's financial secretary, is a fruit selector. The stocky Caribbean woman picks the good banana bunches from a tank filled with chlorinated water and places the fruit on the conveyor belt. Asked about conditions, Auria responds with an infectious laugh:

> The company has introduced environmental certification that has brought improvements and drawbacks. Every three months when the auditors come I have to smile, since the managers panic and rush around to make sure we are in compliance. They instruct us not to speak to the certifiers, but some of us do anyway. They have reduced the chlorine levels and they make us use gloves and wear hairnets. However, the chlorine still makes our eyes water and our hands red.

Auria's wages have also remained stagnant while living costs have increased. For Auria, a fairer system would give all workers, and especially women, a greater control over their working conditions (interview with Vargas Castañeda 2004).

Jiménez, Enrique, and Auria endure these arduous conditions because banana cultivation offers an important source of employment in tropical locations where jobs are scarce. As members of allied unions, however, they demand that banana workers be treated more fairly!

Regional Union Leaders

While the big three global banana firms, Chiquita, Dole, and Del Monte, have admittedly improved certain environmental and social practices, labor organizers still find it an uphill battle to form unions, negotiate contracts, and implement agreements on their plantations, and they find it even more problematic to deal with the independent suppliers of these companies. At times contracted workers also complain about smaller banana cooperatives. These leaders call for a kind of fairness that makes sure that *all* banana workers have the freedom to organize without recrimination and to gain legal agreements over wages and working conditions with their employers.

A large man in his fifties with big wrinkles under his eyes who can be found sweeping the third-floor union office on Fifth Avenue in San José, Costa Rica, Ramón Barrantes is a regional leader of banana workers (fig. 1.3). After years in Del Monte fields, he learned how to organize, to petition courts, to lobby legislators, to make small talk with international visitors, and to write grants for social and environmental training. His Monday routine is answering phone calls and faxes; attending union strategy sessions, company meetings, or legislative hearings; and consulting with attorney Warren Flores and several other assistants about particular cases of fired banana workers or impending changes at certain plantations. On Wednesday or Thursday, Ramón walks the five blocks to Terminal Sur to catch several busses through the Costa Rican rain forest, down toward the Atlantic coast, into Sarapiquí, and into the headquarters where members of his regional union gather for training sessions. There, Ramón helps coordinate educational workshops addressing subjects ranging from support for Nicaraguan refugees, to the effects of paraquat, to the new law

CHAPTER 1

FIGURE 1.3. Union leader Ramón Barrantes at local headquarters, Sarapiquí, Costa Rica. (Photo by author)

that mandates accommodations for nursing mothers. At night he takes another local bus back to his modest home on a plantation's outskirts.

Ramón is clear about what would constitute a fair banana. His union members would gain their own voice in deliberations over wages and working conditions. Otherwise, he remains skeptical: "In Costa Rica it is called '*comercio justo*.' Our conditions are bad. We want to know the criteria for selling such bananas; we suspect that the claimed 'worker committees' which 'fair trade' producers have to set up to comply with global guidelines are really created for their own purposes" (interview with Barrantes Cascante 2004).

NGO Activists

The views of smallholders, workers, and union leaders have been persuasive for activist NGOs or alternative trade organizations that support fairer trade, like Oxfam and Equal Exchange. In 2006, Jonathan Rosenthal, a cofounder of Equal Exchange, agreed to manage OkéUSA, a new fair-trade banana import company. Influenced by the Gandhian principles of *Sarvodya* in Sri Lanka and by the Sandinista Revolution in Nicaragua, Jonathan believes trade fairness is "a simple, concrete and graspable idea. It fits into the consumerist culture and yet also challenges the unconscious acts of consumption we are all urged to make every day. Fair trade gives people a positive way to make a difference." Familiar with the various views of banana activists in the United States and Europe, Jonathan wants to make sure that fair trade is not a top-down, northern-imposed program. Southern producers must be its guide. Yet he hopes that a safe space for dialogue will reveal common interests that bring growers and consumers together. Jonathan recognizes the concerns of unions, yet like many NGO activists, he remains apprehensive about including plantation bananas in the fair-trade mix (interview with Rosenthal 2007).

Theory and Method of Study

Despite embodying divergent meanings of fair trade, these examples all suggest an underlying desire for the equitable treatment of banana producers. My hypothesis predicts a common agreement that those who grow and pack bananas ought to receive a livable wage in a healthy environment. A corollary hypothesis is that the unified action of all these parties could reach this objective. Specifically, in this study I examine whether a farmer-worker-consumer alliance can rally to promote a Fair Trade label in the banana sector as a stamp that conveys both consumer commitment and the achievement of healthy livelihoods for growers.

To pursue such a thesis, I accept in this study that the meaning one assigns to fair trade is very much linked to one's location within the banana-production system. Thus, any reconciliation or common understanding of fairness will depend on the extent to which banana producers, consumers, and supporters can connect their common interests. Development theory and social-movement theory each play a role in clarifying and researching the potential convergence of these interests. While I accept capitalist development and trade as the prevailing model, I do not assume the model's permanence. Rather, I pursue a theory of development that extends beyond modernization and dependency to stress the importance of human agency. Significant transformation requires a major social movement that coalesces the disparate interests and agendas of small farmers, unions, consumers, and dedicated NGOs.

As I elaborate in chapter 5, successful movements require at least three conditions: a structural opportunity, a mobilizing network, and a shared identity. If small banana growers, unionized workers, and smart customers wish to build a unified alliance, they first need a political chance to do so. They must share necessary resources to act on their vision. Finally, to make it all work, they must sustain the view of themselves as participants in a fair-trade project.

While such theories guide us, we must also understand the context of their application. In chapter 2, I consider corporate structures. In chapter 3, I examine how fairer trading emerged as a precursor to what later became the Fair Trade movement. After chapters 4 and 5, which provide the theoretical basis for unified action, I consider structural realities of environmental limits (chap. 6) and trade regimes (chap. 7).

In the latter part of the book, I evaluate specific bases for unified action, beginning with unions (chap. 8), followed by smallholders (chap. 9). Unions pursued a Global Framework Accord with Chiquita, which I discuss in chapter 10. Companies also sought independent certifications, which I evaluate in chapter 11. In chapter 12, I directly examine the important question of Fair Trade and free association. In chapter 13, I specifically consider Dole and Fair Trade, and in chapter 14, I summarize recent trade developments. In chapter 15, I draw together my conclusions.

Data collected for each stage of this exploratory study primarily represent a mixture of qualitative techniques—participatory observations, in-depth interviews, specific case studies, and secondary data collected by others. The approaches I rely on include ethnographic and ethnomethodological elements, narrative and discourse analysis, and historical-comparative methods applied over more than ten years of research.[5] My primary interactions have been with banana workers and local labor officials in Central America, and with a notable number of company officials, certification monitors, and NGO advocates. Transcriptions and translations of their words have been my own. My information on smallholders significantly relies on the observations, narratives, and translations of others.

Summary

Northern consumers are paying more attention to the working conditions faced by those who grow their food, first coffee, and now bananas. As consumers seek a fairer system, they are discovering that fairness has different meanings for the various participants involved in banana production. My thesis in this study is that even within the current "neoliberal" trading system, there can be a convergence of meanings that results in an active, small farmer-worker-consumer alliance to achieve banana fairness. Although divisions have occurred and pitfalls exist, the Fair Trade label represents this opportunity for collaboration. Guided by development and social-movement theory, I test the thesis by considering how components of the alliance historically evolved, how their roles are perceived, and what their likely interactions will be. First, I consider how corporations came to dominate the banana story.

CHAPTER 2

Corporate Banana Structures

ALTHOUGH GEORGE DEFREITAS and Nioka Abbott hope they can gain an adequate livelihood from growing bananas, Jiménez Guerra, Auria Vargas, and the others involved with corporate plantations are less confident. When corporations hold near total control over banana production and trade, it is a major structural impediment to working people earning a living.

The past offers lessons in how these structures of power inhibit workers and small farmers. Company domination increased at the end of the nineteenth century and expanded for the next hundred years. George, Jiménez, and Auria Vargas's ancestors in the Caribbean and Latin America often had to deal with corporate systems that undermined both small farmers and workers. In this chapter I first consider the growth of large plantation companies that resisted worker unions. Second, I look at how companies gained control over the small farmers and pressured governments to remove their protections. Third, I show how a major change in banana production that began in the 1970s made transnational firms more dependent on local landowners and independent producers, and more recently, on supermarkets. These historical developments created structural opportunities for a convergence around "fair bananas."

Corporate Control

Historically, large corporations quickly took control of commercial banana production. Noticing the popularity of local bananas, boat captains brought some back to northern countries. Shipping companies, quickly assessing the bananas' sales potential, carved plantations from lowland rain forests in Central America, Colombia, and Ecuador, and promoted exports from the Caribbean. The U.S.–based United Fruit/Chiquita Company led the pack of transnational corporations (TNCs) in attacking the tropical landscape to set up growing operations. European firms such as Fyffes and, later, Geest became the appointed agents for smaller Caribbean producers.[1]

United Fruit's origins began after 1871, when the Costa Rican government contracted with Minor Keith, a youthful engineer from Brooklyn, New York, to design and construct a railroad to the Pacific from its capital, San José. Keith cleared hefty swaths of land along his railroad right-of-way to use for feeding the crews and soon discovered that he could plant and ship bananas. Keith assessed the fruit's commercial potential, yet won few accolades for his treatment of workers. Many faced sickness and low pay. Even though this new source of wealth brought jobs to manual laborers—as well as to administrators—five thousand died building his railroad (Stewart 1964). In 1899, Keith officially merged his holdings with those of the Boston Fruit Company to form United Fruit, combining huge plantations in Costa Rica, Honduras, Guatemala, Panama, and Cuba with shipping resources from the Dominican Republic and Jamaica.[2]

United Fruit became the largest banana producer among a hundred competitors, including the predecessor of Standard Fruit/Dole, started by four Italian brothers. The Viccaro brothers began purchasing the yellow-green produce from family friends in Honduras and transporting it to New Orleans. By paying higher prices and offering credit even when they were not able to load shipments, they won a reputation for honesty (Karnes, 1978). To prevent such bothersome competition, Keith bought out the Viccaros; a 1908 U.S. congressional inquiry forced divestiture, however (Myers 2004, 43). The brothers reorganized in the 1920s as the Standard Fruit and Steamship Company, which Castle and Cooke, owner of the Dole brand, acquired forty years later.

United also faced another reputable competitor in Samuel Zemurray. In 1910, this Russian fruit trader founded Cuyamel Fruit, which likewise shipped bananas from Honduras to New Orleans. Competition in Honduras aided smallholders in forming a class of their own, and for several decades savvy growers jockeyed their advantage among larger fleet owners (Euraque 1996; Soluri 2005). United Fruit shamelessly expanded its export and transport activities to stifle such possibilities, gaining a reputation as "El Pulpo, the Octopus." By 1915, it owned a hundred ships for banana transport, soon known as "The Great White Fleet."[3] By 1927, United accounted for half the world's banana shipments of nearly 100 million bunches, five times what it had exported in 1900. By 1933, the giant controlled 63 percent of the U.S. market (Kepner and Soothill 1935;

Jenkins 2000, 20). Owing to United Fruit's tense competition with Samuel Zemurray, in 1929 the U.S. Department of State encouraged United Fruit to buy out the Russian. This made Zemurray its top shareholder. When the Depression hit a year later and UF stock fell to 10 percent of its value, Zemurray took control of the United Fruit board, lowered prices to independent growers, and cut "surplus" workers. As company stock doubled in value, Zemurray reputedly improved United Fruit's relations with local governments (Jenkins 2000, 28).[4]

Nevertheless, United Fruit's shenanigans persisted. As banana exports provided Central American nations with increased foreign exchange, El Pulpo gained even more political clout. Company officials won beneficial concessions in taxes, lands, and imports that left little revenue for national coffers. With several million acres throughout Central America, the TNC also became the largest corporate landholder.[5] In Guatemala alone, El Pulpo controlled 20 percent of the land, much of it uncultivated. Having faced a difficult bout with Panama disease in the 1920s (see chap. 6), United Fruit cited potential blight or bad weather as its rationale for holding extensive property. In reality, United Fruit created a vast system of sprawling fields and clusters of workers dotted with wealthy enclaves to fortify market control. It was the largest employer in Latin America, and when it sniffed resistance, it imported laborers from the Caribbean whom it considered docile and adaptable.

When workers collectively reacted, the TNC unleashed its terrifying power, bolstered by local politicians. In one instance in 1930, United Fruit ordered Colombian authorities to kill hundreds of strikers (see chap. 8). By 1935, United's cozy links to state functionaries enabled most Central American nations to merit their characterization as "banana republics." Through its first sixty years, United Fruit had thus become the symbol of corporate and imperialist control (Kepner 1936; Dosal 1993; Langley and Schoonover 1995; Bourgeois 2003; Bucheli 2005; Chapman 2008), while workers sought stronger unions to challenge El Pulpo's unbridled exploitation.

Smaller-Scale Control

Heralding the need for fairer trade, earlier generations of Caribbean farmers comparable in circumstance to George DeFreitas and Nioka

Abbott faced an alternate model of corporate control. Although they grew crops less intensively, mixing bananas with other plantings, by the 1860s, they had begun to export small quantities of fruit. Late in the nineteenth century, investors from Britain, France, and Spain who had learned of Minor Keith's success grew determined to exploit shipments from their own colonies. They strategized about how to transport the fruit quickly, before it spoiled. After the United Fruit merger in 1899, Jamaica especially feared the company would go elsewhere for supplies. Colonial governments determined they could champion smallholder production by working with firms committed to import fruit from their region. As Laura Raynolds and Douglas Murray (1998, 13) explain, Britain and France in particular "made bananas a central vehicle for colonial rule . . . linking the Caribbean to Europe" via banana shippers and state administrators who set up grower associations that pulled together thousands of tiny producers. Fyffes, originally a tea importer, became a shipper of Canary Island bananas in 1890 and expanded to Jamaica in 1901, aided by subsidies and improvements in boat refrigeration (Stiffler and Moberg 2003, 11). However to fill shipments, Fyffes soon teamed up with United Fruit, which fully acquired Fyffes by 1914 and took advantage of British subsidies to Jamaica until the 1930s.[6]

Despite El Pulpo's tricks, several European nations persisted in seeking an alternative banana-production system for their colonies. After the 1940s, London officials encouraged peasant producers in the Windward Islands, especially St. Lucia, St. Vincent, and Dominica, to cultivate and market the fruit through state banana associations that dealt exclusively with Geest, the other prominent U.K. shipping firm.[7] French administrators created similar smallholder systems in Martinique and Guadaloupe that exported via Compagnie Generale Maritime. Spain and Portugal set up cultivation systems in the Canary and Madera islands. Owing to these arrangements, the United Kingdom, France, and several other European countries willingly paid a premium for bananas from the ACP region, and these relationships gradually became codified in the foreign policies of each country.[8] Smallholders fared better under this system than they did in their dealings with United Fruit. They could depend on a more stable market and price for their banana harvests. Nevertheless, they still faced many vulnerabilities in their arrangements with the Geest company.

United Fruit's Radical Challenge

Meanwhile, United Fruit kept restive governments at bay by surrendering a sliver of its vast wealth to its increasingly unionized workers. But when United Fruit workers in Guatemala demanded improvements after the 1944 election of Juan José Arevalo and subsequent labor reforms, it suspended this practice. In 1953, Guatemala nationalized a considerable portion of United Fruit's unused land for needed domestic food production.[9] The government agreed to compensate the company for its assessed taxable value, but after subtracting the fruit giant's excess profits, it owed United very little.[10] While the country's nationalist leaders knew El Pulpo would not welcome the decision, they did not anticipate the ferocity of its reaction or the consequences of the banana elite's U.S. political linkages. United Fruit's Boston law firm included Secretary of State John Foster Dulles. His brother Alan, who was director of the CIA, served on the United board, and the Dulles family quickly assumed a prominent role in promoting a CIA-backed takeover of Guatemala in 1954 (see Immerman 1982; Kinzer and Schlesinger 1982).[11] The company placed its ships and communication system at the disposal of the U.S. Navy, as a land force entered Guatemala from Honduras to depose the Guatemalan president and install a puppet regime. U.S. banana interests similarly inspired official interventions in Honduras, Nicaragua, and elsewhere, although they encountered some resistance in Colombia and Ecuador (Striffler and Moberg 2003).

Throughout the hemisphere, United Fruit declared victory, retaining its domination over banana shipments, including those of Fyffes, which it controlled. However, along with unpredictable weather conditions, it could not suppress all nationalistic reactions or oversee production in former French, Spanish, and British colonies such as the Windward Islands. In addition, its competitor Standard Fruit had expanded from its Honduran base to set up export arrangements in Nicaragua, Cuba, Jamaica, and Mexico. In 1964, the diversified sugar company Castle and Cooke began marketing Standard Fruit bananas under the Dole label. By 1970, Standard Fruit/Dole had joined United Fruit as a dominant force in world banana sales; by 1980, Dole surpassed United Fruit's banana shipments to the United States.

Internal struggles caused El Pulpo additional losses. In 1969, corporate mogul Eli Black suddenly bought out the banana giant and merged it with AMK Corporation, owner of Morrell meats and other products, under the logo United Brands. About this time, the Federal Trade Commission, which had filed an antitrust suit against the company in 1954, required United Fruit to divest itself of about 10 percent of its productive areas and restructure its operations. Between 1954 and 1984 United substantially cut its cultivated lands in Central America from 135,000 acres to 35,000 acres (Stover and Simmons 1987, 431).[12] One beneficiary was Del Monte, a large vegetable producer that took over most United Fruit land in Guatemala (and expanded in Costa Rica)(See fig. 2.1).[13] Then, in 1974, just as the isthmus countries were forming the Union of Banana Exporting Countries to demand a 2.5¢/lb. banana tax (more than triple the going rate), Hurricane Fifi inundated the Central American coast, flooding 70 percent of United's banana plantings in Honduras. To make the tax increase disappear, United Brands and Eli Black made the mistake of offering a $2.5 million bribe to General Oswaldo Lopez Arellano, president of Honduras. When the U.S. Securities and Exchange Commission verified the transaction, it halted trade of the company's stock. Considering the losses, United Brands Chairman Eli Black defenestrated himself from his New York office to his death forty-three floors below. Castle & Cooke, also buffeted by trade union militants, considered leaving Honduras, but a military coup d'etat overthrew Oswaldo Lopez. Key appointees of incoming president General Alberto Melgar Castro had legally represented the firm. Turning to this new source for corporate payments, the army guaranteed it labor peace (see chap. 8).

Nevertheless, in 1978, the European Court of Justice reaffirmed the findings of other tribunals that United's market dominance had violated antitrust regulations and that United must downsize. The company began to recover only in 1984, under the leadership of Carl Linder. It sold off nonfood operations and moved its offices to Cincinnati, touting the fresh name "Chiquita Brands." Although it had to divest itself of Fyffes and other properties, Chiquita gradually regained primary leadership of world banana sales, winning a 33 percent stake by 1990, compared to the 22 percent held by Dole. To accomplish recovery, however, Chiquita employed many of the same manipulative strategies it had used in its

FIGURE 2.1. Del Monte banana worker, Guatemala. (Photo by author)

earlier years, bribing governments, undermining union contracts, and skirting environmental laws. By establishing clandestine companies to escape tax, labor, and property regulations, it could apply toxic pesticides without oversight and relocate workers without their assent. In 1998, the *Cincinnati Enquirer* published the results of a yearlong study on the company's questionable business practices.[14] Because some e-mail information was obtained illegally, the paper subsequently retracted the series, but no one questioned its reported facts or claimed that banana workers or small producers were any better off.

The Colombia Fiasco

In one of its more notorious misadventures, Chiquita also became associated with drug activities in Colombia. Because of internal conflicts in the Urabá banana-growing region within Antioquia, beginning in 1989, the company began paying Colombian guerrillas such as the eighteen-thousand-strong Revolutionary Armed Forces of Colombia (FARC) to protect banana-harvesting operations (see chap. 8). In the early 1990s, when Álvaro Uribe, president of Colombia (2002–) was governor of Antioquia, the Colombian military broadened its antiguerrilla campaign. To perform surveillance and security, the military created a network of rural cooperatives known as Convivir that aroused U.S. suspicions.[15] Uribe promoted Convivir and asked the government to arm its members to aid in capturing guerillas.[16] Over the decade, Convivir developed strong ties with the United Self-Defense Forces of Colombia (Audodefensas Unidas de Colombia [AUC]), a paramilitary force of 15,000 in the region, headed by Carlos Castaño. Regular police presence dwindled as Convivir collected funds from local businesses to support its illegal activities. Castaño approached Chiquita's Banadex affiliate with the AUC plan to "drive the FARC out of Urabá," warning that "failure to make payments could result in physical harm to Banadex personnel and property" (Chiquita Brands 2007). Starting in 1997, and for the next eight years, the company registered on Convivir's books payments totaling $1.7 million, including fifty payments made after 2001 (Evans 2007).[17]

Protection may have come in other ways as well.[18] During 1997, European narcotics agents confiscated more than a ton of near-pure cocaine from at least seven Chiquita boats, with a total value of $33 million. In October, a Chiquita ship from Santa Marta, Colombia, found itself under scrutiny by Belgian authorities in Antwerp, who discovered an estimated $18 million wholesale worth of cocaine hidden behind the insulated wall of a banana container. Company officials, embarrassed by this breach of security, issued assurance that Chiquita was cooperating with the U.S. Customs Service regarding the prevention of drug smuggling on company-operated vessels. However, Colombian sources charged that Chiquita's payments to Colombian paramilitaries were much higher than the company voluntarily acknowledged and included arrangements for exporting drugs and importing guns for paramilitary purposes.[19]

It is irrefutable that Convivir/AUC murdered thousands of innocent people. It operated as a vigilante force,[20] but also worked with the Colombian military.[21] Between 1997 and 2004, the Colombian government listed twenty-two AUC massacres, for example, in Rio Sucio, in December 1996; thirty in Marjripán in July 1997; and fourteen in La Horqueta in November.[22] *Christian Science Monitor* reporter Sibylla Brodzinsky described one particularly bloody day in January 1999 when "14 people were murdered in a killing spree that spread throughout the banana belt's four municipalities" (2007).

By no later than 2000, senior Chiquita officials knew that Convivir and AUC were violent organizations but still approved the payments. In September 2001, after the United States had designated the AUC a terrorist organization, Chiquita persisted in making nineteen more payments, totaling $825,000. It took twenty more months to disclose them. As surprising, the protection bribery continued for an additional year, despite warnings from Chiquita attorneys and knowledge of the practice on the part of the U.S. government. Banadex was Chiquita's most profitable affiliate, and this may explain why the company took so many months to arrange its sale to local independent producers—completed in 2004 for $43.5 million.[23]

In 2007, to compensate for illegalities in supporting a terrorist organization, Chiquita accepted a plea bargain to pay $25 million to the U.S. government over five years. President Uribe and Colombian officials threatened to indict corporate officials (Romero 2007), although given their own involvement with drugs and paramilitaries, this prospect appeared unlikely. Nevertheless, families of affected victims did pursue lawsuits in U.S. courts under the Alien Tort Claims Act, claiming that for profit, Chiquita was indifferent to murder (ACILS 2006; CBS 60 Minutes 2008).

During all these machinations in the 1990s, Chiquita's competitors, who were also accused of making protection payments, expanded sales.[24] Chief rival Dole legally separated from Castle and Cooke in 1995, the latter retaining real estate and resort interests. Adopting a shrewd and

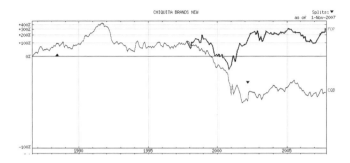

FIGURE 2.2. Chiquita and Del Monte stock performance, 1988–2005. (Graph by Morningstar)

flexible approach, Dole successfully promoted fresh selections of fruits and salad materials in the European market (see chap. 13). Del Monte, which as the world's leading pineapple producer carried its own history of "canned imperialism" and labor exploitation (Burbach and Flynn 1980, 164–219), was purchased by R.J. Reynolds Tobacco Company in 1979. In 1989, the latter spun it off as Fresh Del Monte Produce (FDP) to Polly Peck, which soon went bankrupt. The Grupo Empresarial Agricola Mexicano, headed by entrepreneur Carlos Cabal, bought the firm in 1994, but in attempting to reconsolidate Del Monte, Cabal became implicated in an illegal internal loan scheme that elicited governmental intervention. The Abu Ghazaleh family, from the United Arab Emirates, then purchased FDP majority interest through its IAT holding company and set up headquarters in the Cayman Islands, with offices in Coral Gables, Florida. Tapping supplies from its Chilean fruit affiliate, United Trade Company, FDP expanded its higher-margin produce along with promoting "new formats" for current produce. It acquired U.S. produce distributors in Baltimore and Chicago, and a Belgian marketing company to aid European growth.[25] Already a major purchaser in the Philippines, FDP began sourcing bananas from Brazil and Cameroon. Over the next decade, Del Monte stock strongly outperformed that of Chiquita, as shown in figure 2.2. At century's end Chiquita still ranked first in the global market, holding a 26 percent share, but this represented a decline from 34 percent in 1992 (table 2.1). Conditions were ripe for major changes in TNC strategies.

TABLE 2.1. TNC Positions in the Global Market, 1999, 2004, 2007

	Year	Sales $000mn	Percentage bananas	Income $000mn	Percentage world sales	Percentage U.S. sales	Change world mkt.	Permanent employees
Chiquita	1999	2,556	50	(58)	26		−8	36,000
	2004	3,071	55.4	55	22	25	−4	24,000
	2007	4,663	43.1	(49)	29	17	+5	24,000
Dole	1999	5,060	35	48	35		+4	36,000
	2004	5,316	25.9	61	23	34	−12	36,000
	2007	6,931	26	(58)	26	36	+3	45,000
Fresh Del Monte	1999	1,743	40	57	16		+1	
	2004	2,906	31	139.2	15	20	−1	26,000
	2007	3,366	35.6	179.8	17	15	+2	35,000
Fyffes (Turbana)	1999	1,783	23	63	7		=	4,600
	2004	2,599	17	82.3	11		+4	
	2007[a]	775	25	14.2	3		−8	4,500
Noboa Ecuador (Bonita)	1999				10		+3	
	2004	400			7		−3	2,700
	2007	400+			6		−1	

Sources: Estimates drawn from data in *Fruit Trop* France 1999; van de Kasteele and van der Stichele (2005, 15); company annual reports; author calculations. In 2006 most firms gained 1 percent of world share; Fyffes dropped to 8 percent.
[a]2007 Fyffes's figures represent only its Group Revenue from Continuing Operations.

New Strategies: Sharing the Risk

In Central America during the 1990s, banana TNCs retained their ability to substantially influence state policies. They were less successful in Ecuador and Colombia, however, where peasant and worker reactions forced a different approach. Chiquita had begun losing influence in the 1960s, when smallholders embittered by UF contracts in Ecuador persuaded the company to sell its direct operations to cooperatives (Striffler 2003).

In fact, in the 1960s and 1970s, all the major banana companies reduced their risk by shifting crop-growing responsibilities to independent "associate" producers (see Burbach and Flynn 1980; Bucheli 1997; Grossman 1998). These independent producers were usually local landholders. The TNCs would furnish technical advice and inputs, including banana seedlings, equipment, and pesticide applications. Corporate scientists cultivated plantings more resistant to the black sigatoka fungus that attacked banana leaves, and TNC officials routinely inspected the independent packinghouses to assure export quality. While the arrangement appeared mutually beneficial, the local producers had assumed the risks of market swings and bad weather. They became the primary deflectors of ardent nationalist sentiments and nettlesome labor relations, yet gained no control over the marketing process itself. Some began to question why.

During the 1980s and '90s, as the TNCs expanded production acreage and modernized shipping fleets in anticipation of higher Russian, eastern European, and Asian demand for bananas, they found their independent-producer arrangements increasingly valuable. In exchange for independent loyalty, the companies confronted militant unions. To undermine labor contracts in Costa Rica, they introduced, in collaboration with conservative church leaders, Solidarista associations that replaced unions (see chap. 8). To stave off growing consumer concerns about the ecological impact of banana production, the TNCs engaged independents in joint environmental monitoring programs (see chaps. 6 and 11). Nevertheless, their often-clandestine agreements could easily disadvantage the associates as the TNCs quietly adjusted the cadre that rented and/or administered their land.[26]

The 1998 Hurricane

In October 1998, however, the TNCs desperately petitioned their independents for bananas following the devastation wrought by Hurricane Mitch, which smashed into Honduras, Guatemala, and Nicaragua. Chiquita, which lost virtually all of its own Honduran production, turned to its producers in Costa Rica and elsewhere for fruit. Del Monte solicited its contractors in Costa Rica. Dole, which had already begun to expand production in South America, augmented small-producer relationships in Ecuador.[27] This change had an unanticipated impact: the independents in Ecuador suddenly became power brokers.[28] The TNCs had begun absorbing their output in the late 1980s, but in the 1990s, national firms like Noboa and Reybanpac had developed sufficient markets to exploit Ecuador's low labor costs.

Exportadora Bananera Noboa, "part of a conglomerate of 110 companies called Grupo Noboa, privately owned by Ecuador's richest man, billionaire Álvaro Noboa," exported bananas and other fruit under the Bonita brand. It owned seven thousand hectares, sourced from six hundred smaller independent producers, and controlled nearly a third of Ecuador's exports with help from its own shipping line (Perillo 2005).

With assistance from the Ecuadorian government, Noboa and Reybanpac eliminated most unions. At the same time that Chiquita and Dole were paying banana workers in Colombia, Costa Rica, and Panama $10 a day or more plus many benefits, these national firms gave Ecuadorian workers less than $5 a day without benefits, and often kept them in "temporary worker status" that offered no legal rights.[29] The large Ecuadorian independents also paid little heed to environmental regulations or smaller producers, for whom they proved to be unreliable distributors. Yet with more up-to-date equipment for planting and shipping, the larger firms could export bananas at roughly one-fourth the price of almost anybody else in Latin America. As these factors gave Ecuadorian producers a competitive edge, their increased output led to a worldwide glut in bananas.

The independents then moved to take advantage of this glut. In addition to negotiating more beneficial TNC arrangements, Noboa and others bypassed their marketing dependence, invaded TNC-controlled sales outlets, and forged direct agreements with U.S. and European supermarkets.

At first, Chiquita resisted the independents' newfound strength, but Dole jockeyed to elicit deals. The pineapple giant had gained a larger share of the European market by forging alliances with European companies with import licenses, such as Fyffes, which had just created a joint venture to take over Geest (See chap. 9). Dole then purchased UBESA (Unión De Bananeros Ecuatorianos, SA [Union of Ecuadorian Banana Producers]), the third-largest Ecuadorian independent, to augment banana shipments, as it terminated many of its better-paid workers in Central America. Chiquita and Del Monte then followed Dole's example, increasing Ecuadorian purchases and laying off elsewhere, although not to the same degree.[30] In Costa Rica independents such as Caribana also expanded their power.[31]

Supermarkets Become Major Players

The separate marketing systems developed by independents caused the TNCs to lose control over falling banana wholesale prices. Yet mega food stores "remained immune from the crisis" by setting the terms of retail trade (Perillo 2000). Between 1997 and 2000, the top five chains increased their market share from 24 to 42 percent. Supermarkets relished how independents had broken away from TNC control. In spite of the global oversupply, the fruit remained the retailers' most profitable item. It carried a 40 percent markup, and garnered 2 percent of gross supermarket sales. Through mergers and acquisitions, mega stores had consolidated the consumer market, so now they were taking 80–90 percent of supply chain profits and were in an excellent position to arrange direct contracts with the independent banana producers.[32] The major chains sought national suppliers that could assure a significant quantity and assortment of quality fresh produce. They pursued "preferred suppliers" that would design and provision shelf layout, ad promotions, and planning for sales events (van de Kasteele and van der Stichele 2005). Some chains demanded price discounts; others looked for information systems that promised a sustained inventory. To become preferred suppliers, TNCs had to improve their distribution and tracing processes by carefully monitoring health and chain-management issues and by further rationalizing production and distribution.[33] Chiquita found it had to bid against the independents as well as Dole to deliver the best banana deal to U.S. firms like Kroger, Safeway, A&P, and Wal-Mart, a firm that

TABLE 2.2. Top Five Grocery Retailers, USA, 1997, 2000, 2004, 2008

1997	2000	2004	2008
Kroger Co.	Kroger Co.	Wal-Mart	Wal-Mart
Safeway	Wal-Mart	Kroger Co.	Kroger Co.
American Stores	Albertson's	Albertson's	Costco
Albertson's	Safeway	Safeway	Supervalue
Ahold USA	Ahold USA	Ahold USA	Safeway

Source: Progressive Grocer, 1997, 2000, 2004; Supermarket News, 2008.

now competed with Carrefour in France and Ahold in the Netherlands for retail domination (table 2.2).

In Europe, companies like Spain's Iberga chain, which already held the franchise on importing ACP bananas, freely chose among suppliers. ASDA (a Wal-Mart-owned U.K. discount food chain founded by the 1965 merger of Asquith and Associated Dairy) bought part of Geest and arranged directly with Caribana in Costa Rica. "Due to the Wal-Mart effect, supermarkets are increasingly contracting for produce delivery for six months or a year, locking in an exclusive arrangement" (interview with Rosenthal 2007). The TNCs met the challenge by buying fruit-ripening and import/distribution companies in Europe, and by expanding product lines to include such items as pineapples and tropical fruits, cut vegetables, prepared salads, and packaged, ready-to-eat produce to which they gave strong brand promotion. They offered longer supply contracts, and management and labeling assistance.[34]

Competitive supermarket strategies affected supermarket workers as well as banana producers abroad. Action Aid documented how U.K. "buying practices contribute to poverty wages, dangerous conditions, long hours and insecure jobs for women working on farms and in factories across the developing world," just as supermarket price wars forced Costa Rican women on supplier plantations "out of regular work into casual piece-rate jobs for lower wages" (Action Aid 2007).

The rise of independent producers and supermarkets caused a decline in UF/Chiquita's market control; it also created fresh opportunities. Since TNCs knew more than independents or supermarkets about producing ecologically healthy bananas, Chiquita and Dole attempted to outflank Noboa and other independents who had little taste for bettering environmental, smallholder, or worker conditions. They expressed interest in fairer trade. Yet the public also recalled the TNCs' past disregard for nature and employees. Farmers like George DeFreitas sought a new strategy to draw their own market attention.

A Structural Change

In the middle decades of the twentieth century, two corporate-controlled structures of power ruled banana production: the TNC United Fruit, and its lesser competitor, Standard Fruit, dominated Latin banana production and trade, with few benefits to the countries and communities where they operated. In the Caribbean, Fyffes and then Geest imported from British colonies, while smaller shipping firms handled fruit from elsewhere under a somewhat more equitable system. Despite the drawbacks of TNC domination, unionized workers and small-scale banana producers both won certain protections. In the 1990s, however, overexpansion and adverse weather conditions put the two systems in jeopardy. Touting the benefits of free trade, global leaders slashed preferential guarantees. Supermarkets and independent producers established direct production arrangements. As structures degraded their lives, producers on the ground wrestled with new perceptions. Would they instigate a fairer trade program? To this question, I now turn.

CHAPTER 3

The Fair Trade Alternative

IN THE LATE 1980s, campesinos growing coffee in southern Mexico were fed up with low market prices that did not even cover their cost of production. Inventing their own version of fairer trade, they invited northern residents to show solidarity by purchasing coffee specifically displaying a Fair Trade label. Their idea won over more small-producer co-ops in the South and soon caught the attention of specialty outlets in Europe and the United States. Realizing that small growers generally used fewer pesticides and more cover shade, niche coffee shops began touting Fair Trade purchases as a way for consumers to support farmer livelihoods and avoid environmentally damaging corporate production.

Then, in the late 1990s, alternative-marketing organizations added bananas to their list of Fair Trade products. The move stimulated a range of new questions. Being quite perishable, bananas had to be cut, shipped, and sold quickly, a daunting challenge for niche operations. In addition, whereas coffee was primarily grown on small and mid-sized plots, 80 percent of export bananas were systematically planted and harvested on extensive acreages. Large production cooperatives often did not provide hired workers with adequate wages or protections. Based on *their* understanding of fairness, unionized banana workers then questioned the accuracy of a Fair Trade label. Some fair traders sought to accommodate corporate-related plantations. Yet when TNC banana shippers asserted that their ecological and social precautions deserved the fair-trade stamp, they enraged smallholder proponents. In this chapter I introduce the debate among banana workers, smallholders, NGOs, and corporate interests over the meaning of fair trade.

Fair-Trade Origins

Intentional promotion of fair trade began about 1950, when the Mennonite-related Ten Thousand Villages and Church of the Brethren's Sales Exchange for Refugee Rehabilitation and Vocation (SERRV International) started importing handmade products from developing countries, and retailing them in its stores. Dutch theology students followed suit, selling items through church and volunteer groups. Alternative-trade groups such as Oxfam, Ganesh Himal Trading Company, Cultural Survival, and Conservation International distributed goods that aided a particular people or purpose, ranging from supporting Tibetan refugees to protecting rain forests. In 1965, Oxfam created a bridge program that brought craft products to England from missionary-sponsored artisans in Africa and the Americas. In 1969, the first World Shop opened, and quickly expanded. While kindred to those demanding a new international economic order that would equalize capitalist trade, the religious groups felt the need to act directly. By the mid-1990s, they had established hundreds of northern retail outlets that offered local crafts and circulated materials promoting southern creativity (see Littrell and Dickson 1997; Grimes and Milgram 2000; Ericson 2006).

In the food sector, indigenous Mexican peasants became the major force for alternative trade. After the Union of Indigenous Cooperatives of the Isthmus Region (Unión de Comunidades Indígenas de la Región del Istmo [UCIRI]), a group of fifty-two villages from Oaxaca, experienced a precipitous drop in coffee prices in the mid-1980s, they hit upon the idea of creating fair-trade sales outlets in the North that offered their coffee straight to consumers, thereby avoiding the bidding manipulations of international shippers and retailers. As their pastor, Netherlands native Rev. Franz VanderHoff,[1] explained, when

> on the urgency of the small Indian farmers of UCIRI, Oaxaca, Mexico, (we) established the Max Havelaar[2] market in Holland we did not have a clue where we would end up. But with the farmers I learned that to protest against exclusion, discrimination, exploitation does not make any sense when you don't make at the same time a good, reasonable and valid proposal. . . . Fair Trade as such is not development, but a tool that creates the proper conditions for small producers to get on the road of their own empowerment for development in their own way on their own terms and with their own goals. (VanderHoff 2005)

Since it was difficult to distribute coffee in Europe without access to quotas controlled by the International Coffee Agreement, VanderHoff worked with the Dutch nongovernmental organization (NGO) Solidaridad to roast and distribute Max Havelaar coffee. The brand's original orientation was to support the way of life of independent local producers by offering them a guaranteed market. Solidaridad believed that consumers would willingly pay a little more if their contribution would directly benefit struggling farmers, so its Havelaar Foundation arranged for retailers to purchase coffee from UCIRI that it marketed under a Fair Trade label (Waridell 2002).[3] Each year, Havelaar made sure all participating producers met specified social and environmental standards. In turn, the Foundation certified that those who grew and picked the coffee gained the benefits. In 1988, Havelaar expanded sales to Belgium, to Switzerland four years later, then to Denmark, Norway, and France.

While Havelaar was the first to "certify" fair trade, a similar effort emerged in the United States in 1986 by Equal Exchange, co-founded by Jonathan Rosenthal, Michael Rozyne, and Rink Dickerson. Equal Exchange offered small farmers an advance for their crop at a fixed price, thereby saving them interest fees "as a postcolonial endeavor. We wanted a business that supported socialist movements and helped people in the United States reconnect to food—its origins, its farmers, its nourishing qualities. We were also inspired by the Sarvodaya Shramadana Movement in Sri Lanka. Our initial product line was Nicaraguan and Front Line State coffees" (interview with Rosenthal 2007). By 1989, besides buying coffee from Nicaragua, Equal Exchange had developed its own organic supply chain in Peru. By the mid-1990s, it was retailing $4 million worth of coffee annually and had become the largest fair-trade food marketer in the United States.

Similar "TransFairs" blossomed in other northern countries. Parallel with Havelaar, they consolidated in 1992 as TransFair International, selling nearly five thousand tons of coffee in Germany, Austria, Italy, Luxemburg, and Japan. The same year, Oxfam and other groups founded the U.K.'s Fair Trade Foundation.

Such success inspired Oxfam America and the Minnesota-based Institute for Agriculture and Trade Policy to advocate a common label that expanded the Havelaar approach in the United States. As a result, TransFair USA, became the nation's official fair-trade certifier. Fair TradeMark Canada also started. Led by Paul Rice, who spent more than a decade working with Nicaraguan coffee co-ops, TransFair USA rapidly promoted outlets for Equal Exchange and others.[4] By century's end, alternative trade in coffee involved close to two hundred small, progressive roasters and importers in the United States and Canada. Specialty retailers like Just Us!, Café Rico, Green Mountain Coffee Roasters, Thanksgiving Coffee, and Aztec Harvests championed the benefits of small-scale production and cultivated close relationships with cooperatives.

The Dilemma of Size

As Fair Trade coffee became increasingly popular, its promoters searched for wider outlets to benefit local producers and make Fair Trade accessible to working-class customers. Larger retailers like Seattle's Best, Sara Lee, Van Houtte, and Starbucks developed a sourcing interest (Warning 2005). Rev. VanderHoff and others argued that, as the movement attempted to ship fairly traded goods to a wider consumer market, a broader plantation and retail approach would benefit more indigenous farmers. This persuaded some retailers that an expanded fair-production system might aid local producers who lacked northern norms for paying "prevailing wages." Yet progressive traders and niche roasters like Deans Beans wondered "why small-scale production and democratic principles of organization should be criteria for coffee growers but not for roasters" (Waridell 2002, 105). "The rich countries where fair-trade coffee is being bought have stronger labor regulations and better social safety nets than most of the countries where coffee is grown," explained Laure Waridell. The author of *Coffee with Pleasure* warned that "Some roasters appear to be using fair trade to shield themselves against criticism from consumers . . . (or) have adopted fair-trade coffee in order not to lose customers rather than as a means of assuming their responsibility toward coffee farmers."

Yet many small-farm advocates remained apprehensive as they witnessed Nestle marketing Fair Trade instant coffee in Britain and Wal-Mart's Sam's Club introducing a Brazilian line of Fair Trade coffee in North America that TransFair certified. Procter and Gamble, the largest U.S. coffee distributor, started purchasing Fair Trade supplies owing to pressure from the green NGO Coop-America. In reaction, Equal Exchange itself questioned fair trade tokenism in which giant companies would sponsor a small product line so that they could claim representation on the moral bandwagon of fairness. According to the alternative-trading co-op, if these huge firms were truly committed, they should offer consumers a variety of fair-trade choices to constitute at least 5 percent of their sales (Equal Exchange 2005). Several discouraged, small U.S. retailers finally set up their own certification separate from Fair Trade (Rogers 2004).

The gauntlet drew responses. In October 2005, Starbucks advertised across the United States that a quarter of its coffee sales would come from Fair Trade (see, e.g., *New York Times*, October 27, 2005). McDonalds contracted with Green Mountain to exclusively provide Fair Trade coffee to its restaurants in northeastern states. According to knowledgeable researchers, TransFair was convincing firms to sell Fair Trade "as a way of reducing activist pressure and enhancing their image" (Raynolds 2007, 65).

Producers battled over their own appropriate scale. Did Fair Trade's decision to increase markets through larger companies represent a major departure from its early vision articulated by Michael Barratt Brown? In his book, *Fair Trade*, Barratt Brown portrayed a remedy to unequal exchange between South and North. The system would be involved *in* the market but also operate *against* it, guaranteeing base prices and regulations that buffered southern producers from unpredictable market swings. It would also transfer suitable production incentives, training, finance, and technology to farmers and workers. Barratt Brown envisioned a certification program that tied southern producers and northern purchasers together in a structure that paralleled corporate trading (1993). Yet the difficulty was that when larger firms "crossed tracks" into the system, they promoted being in the market more than against it, as Aimee Shreck (2002) exemplified in the Caribbean. Laura Raynolds (2007) suggested that TransFair's decision to court corporate partners may have improved local living standards, but it did not empower producers. Given that its real coffee prices declined over the decade, others such as Daniel Jaffee (2007) and Eric Holt-Giménez, Ian Bailey, and Devon Sampson (2007) questioned Fair Trade's economic benefits. Gavin Fridell reflected activist thinking in concluding that while Fair Trade had taught farmers to become more sensitive to certain socio-environmental criteria, it had not notably altered the marketing system or improved northern awareness of global inequities: "Instead of adopting Barratt Brown's vision ... most analysts now depict the network as a project aimed at attaining developmental gains within the existing [neoliberal, capitalist] trading system" (2006, 13).[5]

Fairness Beyond Coffee: Small Farmers Plus?

While struggles raged over the optimal size for coffee producers and retailers, the fair-trade movement devoted fresh attention to tea, cocoa, sugar, and bananas—products that were even more likely to be grown on large plantations. If such commodities were to be fairly traded, certifiers could not avoid grappling with the treatment of workers as well as of small farmers. Volatile pricing and declining terms of trade affected both, and the workers faced additional exploitation by giant companies and contractors. The movement faced another divisive choice: Should fair trade concentrate on small farmers or should it also certify large corporate growers in order to aid plantation workers?

Many local cooperatives and retailers remained fearful that despite their best efforts, large transnational firms would manipulate plantation certifications in ways that jeopardized fair-trade (FT) principles. However, others sought to include them. To resolve the dilemma, various national TransFair and Havelaar constituencies sought common standards.

Since the dilemma occurred as FT organizations were expanding the kinds of products they handled, it accelerated their search for a more standardized approach to product certification. Unification came in stages. In 1989, southern and northern producers and traders set up the International Federation of Alternative Trade (IFAT) to share information and improve livelihoods. The U.K. marketer Traidcraft urged more efficient coordination through a European Fair Trade Association (EFTA) founded in 1989; Oxfam and others followed with the Network of European Worldshops (NEWS!) in 1994. In addition to reconciling differences over scale, the various national TransFair and Max Havelaar labeling groups strained to consistently ascertain claims about fairness. Convening in Bonn, Germany, in 1997, they formed Fairtrade Labeling Organizations International (FLO) to set criteria and monitor producers. Members agreed to common principles and practices for both smallholders and plantations that FLO-approved monitors would test. In 2002, FLO established FLO-CERT as a separate certification and inspection unit. In turn, national labelers then licensed FT importers. Together with IFAT, NEWS!, and EFTA, FLO created FINE as an advocacy unit that defined Fair Trade as a "partnership . . . that seeks greater equity in international trade . . . [and]

contributes to sustainable development by . . . securing the rights of marginalized producers and workers" (Krier 2005, 21). FLO members agreed to a common label, one for Europe and another for North America (fig. 3.1). However, alternative traders who feared that smallholder cooperatives might be placed at a disadvantage did not join this effort.

FLO responded to the critique that Fair Trade was primarily in, not against, the market by offering an alternative approach that helped producers make direct arrangements with retailers. Arrangements were designed not only to cover costs but also to assure small producers a reasonable livelihood. The arrangements included:

- a fair price, usually above market, that conveys sustainable production and an adequate family income
- a special premium for community use
- a long-term contract that provides stability
- a partial prepayment when requested by a producer organization (to enable continued planting, for example)

For FLO to accomplish its purpose, consumers had to choose products that were Fair Trade labeled as a step toward balancing global inequities and fixing the broken bungee "race to the bottom" of southern workers. By 2007, FLO represented more than twenty national initiatives that inspected and certified 580 producer organizations in more than fifty countries. Fair Trade products, including footballs and roses, earned their producers millions of dollars more than they would have if they had been sold on the volatile open market (15 million euros above world-market banana prices in 2006).

In reply to objections, FLO advocates also stated that a trading system that depended on ethical choices of knowledgeable consumers could bring only a limited answer to the growing inequalities between North and South, and to the unchecked ability of transnational corporations to move production to wherever they wish. In his role as a Fair Trade consultant, Jonathan Rosenthal had also questioned its endgame: "Do we want all the goods in the world to be ultimately labeled 'Fair Trade'? At what point does government policy enter, or social standards become legislated?" (interview 2007). In response, most fair traders were clear that additional governmental trade initiatives must redress these inequalities. State agricultural policies must protect land from export agribusiness

FIGURE 3.1. (*a*) FLO-Certified Fair Trade logo used in the United States (used by permission of TransFair USA), and (*b*) FLO-Certified Fair Trade logo used in Europe (used by permission of Fairtrade Labelling Organizations International).

and emphasize production for local food needs.[6] However, the success of these initiatives would ultimately depend on people in both North and South becoming better informed about the way production and trading systems work. Positive involvement with Fair Trade products might help make this happen.

Spokespersons for Banana Fairness

In the mid-1990s, following the lead of alternative traders like BanaFair (Germany) and Gebana (Switzerland), national labeling organizations took an interest in bananas.[7] Proponents emphasized that producer prices had consistently fallen while retail prices remained relatively constant.[8] Although certain European nations had granted protections to bananas from former colonies, others had not; then, in 1993 the entire European Community adopted common import rules for African, Caribbean, and Pacific (ACP) bananas that reduced tariff and quota protections for island countries (see chap. 7).[9] European NGOs urged greater remedies for the social and environmental dilemmas faced by banana-producing communities. Banana Link, in the United Kingdom; BanaFair, in Germany; and others stimulated the formation of the European Banana Action Network (EUROBAN) to lobby for an equitable trading system that guaranteed an adequate income and a protected environment to farmers and banana workers. EUROBAN welcomed the participation of national FLO labelers. The TransFairs and several Havelaars subsequently invited advice from EUROBAN participants in formulating a set of Fair Trade banana criteria. According to Harriet Lamb, FLO banana coordinator at the time, "In some products, the most disadvantaged are landless people working on plantations. . . . Furthermore, produce from such estates is needed to support smallholders' sales. . . . That's why, right from the early days, Fair Trade standards were developed for workers on plantations" (2008, 147). While control via FLO caused some disgruntlement, BanaFair, Gebana, and others became licensees (Pfeiffer 2008). As EUROBAN turned its attention to fair-trade certification, Solidaridad/Havelaar created AgroFair in 1996 as a joint venture with southern cooperatives to market bananas in Europe under the Oké brand.[10] AgroFair promised to be an FT licensee "passionate about the rights of banana workers" (http://www.agrofair.nl). Surveys indicated that 7.5 percent of European consumers would buy FT bananas at a 10 percent premium, which meant 300,000 tons a year.[11]

Environmentalists took up the fair-banana cause. One of the first environmental groups to do so, Foro Emaús, a Costa Rican coalition that originated in the Roman Catholic diocese of Limón, where banana cultivation was the primary economic activity, detailed the horrendous ecological conditions surrounding the fruit's cultivation.[12] European environmentalists warned about planetary deforestation and pesticide contamination. The Rainforest Alliance, a U.S. organization that had begun certifying timber practices, took an interest in Costa Rica following Foro Emáus publicity, and a leaked government report documenting how banana cultivation absorbed three thousand hectares of arable or pristine land between 1990 and 1992 (Taylor and Scharlin 2004, 33).[13] Presentations at the 1992 International Water Tribunal charged Dole with ruining Costa Rican rivers (International Books 1994). Becoming more aware of Chiquita's nefarious treatment of employees and habitat, the Rainforest Alliance invited the major TNC banana companies to consider a fresh approach. Chiquita, the only one to respond, agreed to a pilot project headed by Rainforest representative Chris Wille that conveyed an ECO-OK stamp on bananas from certain Costa Rican plantations. Working with the National Wildlife Federation, Wille then placed ads in children's magazines, asking subscribers to contact banana companies to "make all their plantations ECO-OK." The responses persuaded Chiquita that "we had a problem.... We looked around and the Rainforest Alliance was the best choice to work with us" (Taylor and Scharlin 2004, 36). The program gradually convinced Chiquita to improve its reforestation, recycling, and pesticide application practices and meet certain social standards.

However, EUROBAN members and U.S. NGOs voiced suspicion about the Rainforest program. Its "ECO-OK" stamp of approval appeared to compete with the "Eco" designation reserved for organic food brands. Its social claims of "fair treatment and good conditions for workers" ran counter to reports about worker abuse from Costa Rica. "We thought we were doing a good thing, and the reaction among NGOs caught us off guard," admitted Wille (1999). Yet Chiquita also persisted in using pesticides banned in the North that threatened worker and community health. Rainforest reorganized its environmental verification, replacing ECO-OK with "Better Bananas." By December 1999, Better Bananas had certified all Chiquita farms as dutifully following a rigid checklist of protections; by 2004 the TNC had verified environmental and social compliance (the latter certified by SA8000; see chap. 11) of most of its independent producers (Taylor and Scharlin 2004).

As a result, Chiquita and other TNCs began claiming they should also be candidates for fair trade. Wille and company officials promoted Chiquita as leading the improvements in plantation conditions. Sylvain Cuperlier, corporate responsibility spokesperson for Dole Brands, likewise praised his firm's monitoring practices and organic division (2005). Del Monte also claimed substantial social and environmental enhancements (M. Garcia 2005). The TNCs cited their promotion of trade regulations as an additional rationale. Chiquita especially expressed concern that the loss of markets in Europe that the former tariff/quota system provided was jeopardizing its ability to treat nature and employees fairly (see chap. 7).

EUROBAN and other NGOs remained unimpressed by corporate claims. Small banana producers and plantation unions reported that TNC ecological behavior had somewhat improved, but social conditions exhibited little change.

Further Efforts Toward Fair Bananas

As NGOs evaluated the competing perspectives on fairer trade, they encountered "involved stakeholders" like George DeFreitas, Nioka Abbott, and Juan Quenteña; workers like Jiménez Guerra, and Auria Vargas; and leaders like Ramón Barrantes. The NGOs discovered issues that corporate officials, even those with an environmental and social conscience, did not fully address. Acting quickly, but in hindsight without sufficient southern input, FLO set up a detailed certification and import program that separately checked on the conditions of banana production at worker-owned cooperatives, on the one hand, and at corporate or independent plantations on the other. Over the next few years, the standards for gaining a FLO label became more specific and elaborate.[14]

TABLE 3.1. Fair Trade (FLO) Certification Standards

Category	Key Requirements
Social development	Honor core ILO labor rights
Economic development	Pay above-market price plus group premium
Environmental development	Avoid chemicals wherever possible. Protect forests and recycle wastes
Trade	Honor contracts and prepayments

FLO Standards for Fair Bananas

By fall 2004, most FLO-approved banana producers underwent a careful testing process to make certain that they were truly fair to all parties. They had to complete an extensive questionnaire, receive approbation by the FLO board, undergo an annual FLO-CERT inspection (see http://www.FLOCERT.org for updated criteria), and have their financial and import data cross-checked. The inspection process involved approved monitors, such as AgroFair personnel, who visited production areas, stringent guidelines in hand. They verified compliance in four areas listed in table 3.1: social development, economic development, environmental development, and trade.[15] Each standard established a baseline but also set up progress benchmarks. Baseline producers were officially listed in the Fair Trade Register (for updates, see http://www.fairtrade.net/productstandards.html).

FLO refined the four categories of standards for the two different types of banana production: The first type was small farmers who grew bananas along with their other crops and animals. They avoided large fields, where pests and fungi could easily invade, and often interspersed bananas with other plantings. While small farmers could hire several others to help with their harvest, it was assumed that there were not many additional workers on family farms (a point the unions would challenge; see chap. 12).

The second was plantation production that involved using extensive acreage. Unlike many commercial operations, FLO-certified plantations had to be scrupulous about the environment. They might not be fully organic since tropical areas have a myriad of pest invaders. Nevertheless,

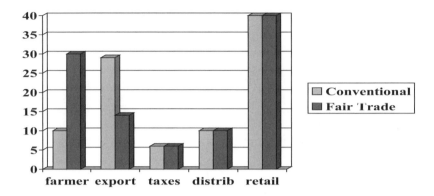

FIGURE 3.2. Percent stakeholder share in Fair Trade/conventional banana prices.

they avoided chemical fertilizers and pesticides wherever possible and had to have a well-developed plan for recycling and disposing of banana wastes. Plantations also faced strict rules for respecting workers based on the International Labor Organization (ILO)'s "core labor standards" (Compa and Diamond 1996; chaps. 7 and 11). Union doubts about plantation compliance became an additional source of contention.

In sum, the FLO label guaranteed a fair price. While inflation could be an issue, the price was ordinarily set above the world-market price, high enough for producer families to feed themselves, send their children to school, and have extra funds (a social premium) to "invest in sustainable development."[16] In Ecuador FT-certified producers generally earned more than $7 a box, compared with less than $5 a box by non–Fair Trade producers. Overall, Fair Trade producers gained 27 percent of the banana retail price, compared to only 7 percent received by conventional banana growers (see fig. 3.2 for price comparisons).[17] FLO rules were also designed to assure a safe environment and good working conditions for plantation workers, including the freedom to join a union.

Fair trade worked if commercial companies were "willing to buy from registered producers on their Fair Trade terms." Retailers had to sign a legal agreement with a national initiative (e.g., TransFair USA) that monitored activities via regular audit reports from importers that detailed their arrangements with sourcing producers (FTF 2004a). Adhering to these social and environmental requirements meant that FLO-labeled bananas cost more, but the stamp also saved purchasers money in bypassing the usual middleman charges.

Consumer Results

Overall, the amount of FT bananas marketed has expanded. As shown in table 3.2, northern sales jumped by an average of 42.5 percent a year between 2001 and 2007 (63 percent 2004/2007), reaching 234 thousand tons. It increased to 234 thousand tons despite higher fuel surcharges (FLO 2007). After Havelaar introduced Fair Trade bananas in the Netherlands, Switzerland, and elsewhere in 1996, it subsequently enjoyed a 100 percent annual increase. Arriving in Denmark in 1997, FLO bananas soon gained 3 percent of the market. Entering Britain in 2000, by 2003 Fair Trade bananas had captured 8 percent of the national market helped by the Fairtrade Foundation's emphasis on consumer awareness. By 2007, a quarter of all bananas purchased in the U.K. were Fair Trade (FLO 2007). Half of the public in the United Kingdom, Belgium, Finland, and Sweden understood the fair-trade concept, and most European supermarkets stocked FT fruit (FTF 2004b; see Krier 2005 for data). With more than 100,000 volunteers, organizations such as Oxfam, Traidcraft, Christian Aid, and CAFOD aided Fair Trade's high annual growth rate.[18] BanaFair distributed in Germany, where the Lidl discount chain became a major FT banana outlet.

As Lamb (2008, 21ff) recounts, marketing setbacks occurred in Germany, Denmark, and elsewhere owing to hurricanes, drought, and mysteriously delayed shipments that caused rotten bananas and order cancellations. "We faced challenges at every step." When she became director of the U.K.'s Fair Trade Foundation however, Lamb "went for scale," approaching Tesco, the nation's largest chain that agreed to promote what customers "are looking for" (33). By 2004, this shift revived sales from the Caribbean, especially Dominica (see chap. 9).

TransFair, in its efforts to rapidly expand the U.S. market, cultivated a business-friendly reputation that created tensions with other NGOs (Raynolds 2007, 71f). Nevertheless, a primary TransFair objective was to help family farmers in developing countries gain direct access to international markets by selling their own harvests (see http://www.Transfair.org). Nearly 10 percent of the U.S. population said they had purchased fair-trade produce, although they often had difficulty finding it (TransFair 2006). In 2004, TransFair began overseeing banana sales. Without notable NGO consultation, the certifier licensed three large and four smaller fruit importers and made special arrangements for Wild Oats, a 102-store grocery chain located in twenty-four states, to purchase four containers per week. Distribution was also slated for Safeway, Stop & Shop, Whole Foods, and Starbucks.

A Fair Trade/Union Agreement

Yet TransFair/FLO's marketing success caused banana unions in Latin America to question its review process. Despite the application of FLO social-development criteria and ILO conventions, despite cooperation with EUROBAN, most plantation bananas that FLO imported into Europe were not produced with union labor. They came from estates in Africa or Suriname and co-op–hired hands in Ecuador, Peru, and Costa Rica who "often total more than 50 percent of the workforce. In addition, when they are short in meeting a shipment, they buy from private nonunion suppliers. Thus, under the guise of being more socially responsible these bananas subvert union bananas," emphasized German Zepeda of the Coordination of Latin American Banana Unions (COLSIBA). "Marketing these 'socially just' bananas may actually violate FLO standards by undercutting worker rights" (Zepeda 2005).

Yet, when faced with the irony of potentially selling nonunion bananas, TransFair became the first certifier to publicize that FLO fruit met social-development criteria and would be union approved. In June 2005, TransFair and COLSIBA signed a special agreement under which plantation bananas would be marketed in the United States with the Fair Trade seal (see chap. 12). In 2006–7, however, various factors slowed the agreement's implementation as well as the overall sales of certified bananas in the United States (table 3.2). First, as Rosenthal noted, the logistics of distributing a perishable commodity to smaller retailers rendered bananas "a difficult market to break into. The fruit must be in the right

TABLE 3.2. National Fair Trade Banana Sales, 2001–07 (metric tons)

Country	2001	2002	2003	2004	2005	2006	2007	Annual percentage change	Market share
Austria		1,775	1,827	1,913	2,804	5,810	7,023	89.0	2.7
Belgium	925	1,314	1,994	3,012	3,536	4,181	5,095	23.1	4.0
Canada				184	239				
Denmark	294	365	609	749	784	1,091	2,307	69.4	0.9
Finland	1,707	2,833	2,497	2,414	5,000	7,306	7,986	76.9	5.0
France	82	696	829	1,773	3,162	4,547	7,208	102.2	
Germany	101	117	131	1,508	3,218	9,760	13,600	267.3	
UK	9,701	11,426	18,182	25,970	40,685	60,276	143,469	150.8	5.5
Ireland			359	501	612	1,058	2,239	115.6	0.5
Italy	20	82	2,039	2,788	3,247	3,929	4,055	15.2	
Japan						141	223		
Luxemburg	168	178	163	203	276	281	223	3.3	
Netherlands	2,303	1,996	2,610	2,736	3,236	3,381	4,313	19.2	
Norway	33	154	302	338	835	780	1,046	69.8	
Sweden	568	586	797	1,169	1,825	2,645	3,647	70.7	0.6
Switzerland	13,170	15,090	18,813	31,766	30,499	27,981	28,336	−4.0	47.0
U.S.A.				3,616	3,919	2,594	3,278	−3.4	
Spain						001	001		
TOTAL	29,072	36,612	51,152	80,640	103,877	135,763	234,112	63.4	

Source: www.fairtrade.net/bananas.html, 2007; Annual percentage change 2004—compared to 2007; Market Share 2004 (Krier 2005).

amount and condition, both in color and volume. This requires local distributors receiving a consistent stream of delivery three to six times a week" (interview with Rosenthal 2007). Second, TransFair failed to gain endorsement from NGO groups like Oxfam America when it launched its banana program at home (Raynolds 2007, 71). Abroad, "TransFair had been working closely with the Council of Latin American Cooperatives (CLAC) for coffee (and some bananas) but then signed with the unions without real CLAC participation. In its haste to recruit the big guys [corporations], it actually exacerbated tensions between the CLAC folks and the unions" (interview with Rosenthal 2007). Third, TransFair did not fully comprehend what union rights entailed, including transparent assurances regarding bargaining, marketing, and control over the FLO social premium.

In 2006, with TransFair's blessing, AgroFair entered the U.S. banana market via a joint venture with Equal Exchange and Red Tomato, aptly named Oké USA. Oké began shipping two to three containers of organic FLO bananas a week to cooperatives and university food services.[19] However, the 2008 economic downturn and short supplies forced Oké to recapitalize under Equal Exchange. Rosenthal moved on to head Just Works Consulting.

Summary

In this chapter I have traced the history of fair trade, from its origins in religious groups seeking just compensation for third world artisans to Mexican campesinos who inspired the idea of a Fair Trade label that would inform northern consumers of the fairness surrounding the production and distribution of an agricultural product. The label created some dilemmas as promoters expanded into products grown by large co-ops and corporations that also staked a fairness claim. It raised the question about whether such expansion threatened Barrett Brown's early characterization of fair trade as a genuine in/out market approach for both South and North.

A FLO-labeling program became one way to address these in/out, small/large dilemmas, yet possibly include plantation production. While not winning universal acceptance, FLO sought to accommodate both smallholder and union concerns that centered around producer treatment. In practical terms, however, although fair-trade consumer interest had accelerated, the expansion of FT banana sales in the United States moved more slowly than had been expected. The reasons this occurred will affect our understanding of a smallholder, worker, consumer alliance in the chapters ahead.

Interesting account of how fair trade has focused mostly on small-commodity producers, but failed to include workers. Also, the main zero-sum game lies exclusively in the first exchange between producers & exporters, but further up the chain, neither distributors or supermkts have taken any hit. This fact highlights the eminently reformist nature of the whole fair trade movement. It amounts roughly to seek the equivalent of minimum wages & improve environ. practices.

CHAPTER 4

Actors for Banana Development

FAIR BANANAS ARE at a crucial juncture. The quandary is whether those actually growing the fruit—both small farmers and unionized workers—can reach a unified understanding with consumers.

Bananas and Development Theory

For consensus about whether growing bananas for market can help a country develop, my historical review of the corporate banana system was not encouraging. The fruit was often planted on land that could have been cultivated for local food production, and southern nations retained only 12 percent of northern retail banana prices (Chambron 2000), suggesting why "banana republics" have remained perpetually underdeveloped.

Yet if we assume the fruit was carefully grown on terrain less suitable for grain and vegetables, what would those who grow bananas like to see happen? I posit in this study that most share an understanding that fair-banana development would bring jobs to more people with sufficient pay for families. Health needs would be met and children would attend school. The natural environment would increase its vibrancy, and community cultural and institutional life would strengthen for future generations. To put it another way, banana development would be *sustainable*.¹

I also anticipate consensus about how such development is achieved. Theorists have often followed one of three trends: those who see development as modernization, that is, "becoming modern"; those who view it as independence, that is, eliminating dependency on other nations; and those who understand it as gaining equality for disenfranchised groups such as indigenous populations, women, and immigrants, all notably vulnerable to economic and environmental change.² Modernization advocates believe that the goal of development should be a progression of "stages" that "takes off" with industrialization and culminates in a consumer society like those of the United States and Europe (Rostow 1971). Instead of promoting small-scale or worker-controlled efforts, modernization theorists tout corporations such as the giant banana firms as being the most successful means for setting up efficient growing and shipping operations.[3] Global financial institutions like the World Bank and the International Monetary Fund have encouraged broad private investment, believing that it rapidly increases jobs and builds necessary infrastructure such as ports, railroads, and even hospitals and schools. The institutions have emphasized that corporate modernization has brought benefits that the local population did not previously enjoy.[4]

For dependency theorists, however, corporate modernization originated in the North, where firms still maintain their wealth and receive state diplomatic and military support. Southern nations that welcomed corporate investment have often remained underdeveloped as they continually court northern companies and governments for markets and sales, in a condition of perpetual slavery (Galeano 1969; Polanyi 2001). As a result, most fair-banana proponents are likely to champion dependency analysis.[5] They often cite the republics of Central America—in which corporations owned vast tracts of land and resources, where local governments dutifully complied with corporate desires, and where the bulk of the population remained abysmally impoverished and uneducated—all typical examples of the efficacy of dependency analysis.

Dependency thinkers therefore urge people to rise in protest and to establish local control with their own systems of production—much as Guatemalans did when they briefly challenged United Fruit/Chiquita in the early 1950s (Immerman 1982). Dependency advocates believe *independence* creates opportunities for small producers and national firms to gain sufficient control over trade to balance the unequal exchange between the North and the South (Furtado 1976; Emmanuel and Pearce 1972; Wallerstein 1976). Southern nations have followed this approach by forming bilateral agreements or creating regional trading systems among themselves.

In recent years, fair traders and scholars have opted for a third approach to development. While retaining a healthy critique of private-sector priorities and northern political influence, these proponents realistically assess the benefits of properly controlled capital investment.

They believe if foreign resources are competitive, countries can negotiate better arrangements. People do have this capability. Historians who have reappraised past conditions in the banana sector have documented earlier resistance to northern domination than scholars have previously portrayed (see, e.g., Euraque 1996; Soluri 2005; Striffler and Moberg 2003).

This third option involves local groups (indigenous, women, environmental activists) taking independent action for their own benefit. Ordinary people grasp the initiative to honor their cultural heritage and still gain a livelihood. The approach signifies the importance of agency: in a globalizing world, sustainable development can successfully emerge from "below," exemplified by small banana holders that seek to control their own production and markets.

In proposing a parallel market structure in 1993, Advocate Michael Barratt Brown promoted Fair Trade as a sensible antidote to both modernization and dependency. It would equalize relations between northern consumers and southern producers, rebalance technical resources, and address additional inequalities that the corporate system could never remedy. Remaining in the market, Fair Trade would also act against it (see chap. 3). By reinforcing the local culture's ecological and collective norms, Fair Trade would eschew modernization and encourage a more balanced, just, and humane commercial bond between small producers and northern customers.

Pursuing the third approach, FLO exposes itself to the risks of which Fridell (2006) warned: that the approach relies too much on voluntary involvement, that it is too accepting of a capitalist marketing system, and that it is too dependent on the voluntary efforts of nongovernmental organizations (NGOs). After his experience with co-ops in Sandinista Nicaragua, Paul Rice, CEO of TransFair USA, acknowledged that Fair Trade is only "a reform movement. I know what revolution is, and this is not revolution."[6] However, most theorists and practitioners understand that no single strategy, including Fair Trade, will accomplish major change by itself. If banana activists could coalesce around the third approach, the movement could still question "business as usual" and work to overcome the total absorption in price.[7] In itself, this shift could shake up the system (Nash 2000; Magdoff, Foster, and Buttel 2000). In any case, as TransFair board president Michael Conroy has pointed out, Fair Trade certainly does not *prevent* broader change. He challenged critics to find a better alternative (2007).

Actors for Development

I turn now to look at some of the possible actors who could usefully adopt the third approach to fair bananas. Table 4.1 summarizes those farmer, worker, and consumer groups:

WINFA

The Windward Islands Farmers Association, the group in which Anton Bowman participates, was created in 1982. Its objective is to promote the social and economic welfare of small farmers in the Caribbean. The organization includes independent producer groups, individual farmers, and agro-processors on four islands (St. Vincent, St. Lucia, Grenada, Dominica) and Martinique. It offers solidarity to farmers seeking livelihoods but also encourages their democratic involvement in building awareness about their dilemmas and challenges.

CLAC

The Coordinadora Latinoamericana y del Caribe de Pequeños Productores en el Mercado Justo (Council of Latin American Cooperatives [CLAC]) grew out of networks of small coffee farmers and beekeepers in 1996. Part of the worldwide Via Campesina movement, CLAC became, in 2002, the voice of registered Fair Trade producers from Latin America and the Caribbean (with a small African delegation), and quickly gained representation on the FLO board. (Unfortunately Carlos Eugenio Vargas from Coopetrabasur in Costa Rica was resistant to unions. See chap. 12.) By 2006, CLAC represented twenty thousand affiliated members in three hundred producer organizations like WINFA from twenty countries (CLAC 2006a). Banana growers represented about 7 percent of CLAC's membership, a size similar to those of producers of honey, juice, cocoa, and fresh fruit.

Unlike WINFA, CLAC retains an absolute fear that "grand TNCs hold enormous power to monopolize the market via powerful, manipulative

TABLE 4.1. Key Fair-Banana Promoters

Farmer Groups and Supporters

WINFA	4,000 farmers in the Windward Islands
CLAC	20,000 small producers in Latin America
Other groups	

Unions

COLSIBA	30,000 workers in Latin America
IUF	12 million food workers worldwide (e.g., 3F Denmark, UFCW, Teamsters)
Other unions, federations	For example, UNITE-HERE, UAW, CGT France

Farmer/Union Supporters

EUROBAN	30 NGO groups from 13 EU countries
BananaLink (United Kingdom)	Co-op that works with unions, farms
Peoples Solidaires (France)	NGO with 80 local southern assns
Other groups	

Union Supporters

Solidarity Center	U.S. labor international support office
USLEAP	Chicago-based NGO coalition
Other groups	STITCH, ILRF, Labour Start, Campaign for Labor Rights, LCLAA

Consumer FT Certifiers/Distributors

Havelaar/Solidaridad/Agrofair	FLO certifier/FT banana importer
TransFair USA	FLO certifier, works with importers
OKE/AgroFair/Equal Exchange	FLO importer, FT distributor
Oxfam (in various countries)	FT marketer via 2,500 World Shops
CTM Altomercato (Italy)	FT marketer via 350 shops
BanaFair (Germany)	FT certifier, works with unions, farmers
Other advocates	

Other groups of interest — Solidarity groups

tools such as anticipated purchases and price speculation." It remains adamant that the Fair Trade designation should be restricted to products from small farmers (CLAC 2006b, Principle #8). It promotes the food sovereignty and self-sufficiency offered by "dynamic ancestral agriculture that emphasizes feeding the community first" (Tiney 2007). Nevertheless, the banana network within CLAC officially supports "unions" and "their accomplishments ... in face of the huge TNCs that exploit the workers" (CLAC 2006b, Principle #4). CLAC hopes to reconcile these conflictive issues soon (interview with Lucio 2007).

COLSIBA

With thirty thousand members, the Coordination of Latin American Banana Unions also serves as a regional body for independent unions from sugar, coffee, flowers, and other agro-industries. SITIGAH (Sindicato de Trabajadores de Plantaciones Agrícolas y Ganaderos de Heredia [Union of Livestock and Agricultural Plantation Workers of Heredia, Costa Rica]), the small banana union headed by Ramón Barrantes, and four other Costa Rican unions are COLSIBA affiliates, along with other unions in Belize, Colombia, Ecuador, Guatemala, Honduras, Nicaragua, Panama, and most recently, Peru. It was COLSIBA that signed the agreement with TransFair USA in 2005.

EUROBAN

Founded in 1994, the European Banana Action Network is a loose grouping of over thirty European NGOs and trade unions in thirteen countries that promotes policies and actions on behalf of small farmers and banana workers. Members include development organizations and alternative distributors that market fairly traded bananas.[8] While its original emphasis was on livelihoods, for both peasants and workers, EUROBAN participants now network together to support a socially, environmentally, and economically sustainable banana industry via alternative trade, worker and environmental rights, and reform of the European banana-import regime. For example Banana Link (United Kingdom), a cooperative founded in 1996, works closely with COLSIBA via its union-to-union project and publishes update bulletins regarding Caribbean farmers.

The development NGO Peuples Solidaires, begun in 1983 in France, has educated its eighty local associations about banana issues.

The IUF

Many COLSIBA affiliates are members of the International Union of Food, Agricultural, Hotel, Restaurant, Catering, Tobacco and Allied Workers' Associations based in Geneva, Switzerland. Originating in 1920, the IUF has become a global federation of 336 trade unions that represent employees in agriculture and plantations, food and beverage manufacturing, hotels, restaurants and catering, and tobacco processing in 120 countries, for a combined membership of over 12 million workers. The IUF network assists federations and unions in campaign organizing, solidarity outreach, educational programs, and research to counteract the growing power of TNCs and the forces of globalization. It has a notable presence in Europe and Asia. Its Danish affiliate 3F, representing 362,000 workers, along with Irish, British, Swedish and French unions, are especially active forming international linkages. In the United States, the IUF includes the United Food and Commercial Workers (UFCW) and some Teamster food processing locals. Since the mid-1990s, the IUF has organized various projects linking small farmers and farm workers in Africa (Hurst 2007, 83).

The Solidarity Center

Formed in 1997 under John Sweeney, leader of the AFL-CIO, the American Center for International Labor Solidarity (ACILS), better known as the Solidarity Center, serves as a primary AFL-CIO Federation vehicle for global union support. In replacing most vestiges of notorious predecessor institutes such as AIFLD,[9] the center represented a fresh attempt to forge a stronger labor movement abroad. The center faces persistent scrutiny from union activists since the bulk of its funding still comes from the U.S. government, but it channels its support to "building operational, pragmatic global unionism" (Coats 1998; see also Gacek 2005). For example, in Ecuador, Solidarity staffers Teresa Casertano and Liz O'Connor worked with FENACLE and Bonita banana workers to gain representation and labor reform (chap. 12).

USLEAP

The U.S. Labor Education in the Americas Project (formerly the U.S. Guatemala Labor Education Project) is an independent, nonprofit organization that supports the basic rights of workers in Latin America, especially those who are employed directly or indirectly by U.S. companies. Founded in 1987, USLEAP has a board of twenty representing a cross-section of labor unions, human rights groups, and religious organizations.[10] Its director, Stephen Coats, emphasizes the importance of coalitional work governed by local union priorities. The NGO has assisted banana workers since 1998. USLEAP has fraternal relations with the Support Team International for Textileras (STITCH), an organization that assists textile- and banana-union women.

Consumer FT Certifiers/Distributors

Besides fair-banana producers, fair traders include national FLO-labeling initiatives, importers (who wholesale, or retail directly), and exclusive FT retailers such as World Shops. Typical labelers are the Fair Trade Foundation (United Kingdom) and TransFairUSA. Importers and distributors include AgroFair, BanaFair, Oké/Equal Exchange and others. Often each importer has active projects in specific banana-producing countries. Germany's BanaFair began importing and distributing fair-trade bananas in 1987 to support unions and improve contracts with small producers. It purchases from the Colombian firm URCOL but also does investigative studies of banana conditions elsewhere. Most FT labelers, importers, and retailers coordinate their activities via the triennial FLO conference, and are suitable candidates for coalitional efforts.

Among retailers, Oxfam's role stands out. Originating during World War II among Quakers and others in Oxford, England, dedicated to famine relief, Oxfam International expanded into a "confederation of 12 organizations working together with over 3000 partners in more than 100 countries to find lasting solutions to poverty, suffering and injustice" (http://www.oxfam.org). Partners such as Oxfam America and Oxfam U.K. promote global public education about social and economic justice as essential for sustainable development. As I noted in chapter 3, Oxfam members have long advocated fairer trade. In Europe, the organization sells FLO-certified goods in 2,500 World Shops. In Belgium, for example, 7,500 Oxfam Werweldwinkels volunteers staff 206 shops, offering products from 106 southern partners. National Oxfam offices in the United States, the United Kingdom, Belgium, Ireland, Canada, Spain, New Zealand, and Australia circulate educational materials and programs to a considerable number of schools, community organizations, and religious groups.

The list includes many others, such as CTM Altromercato in Italy. Started in 1988, it is Europe's second-largest Fair Trade distributor. CTM manages 350 shops through its cooperative network of 130 associations that urge entire towns to become "Fair Trade Communities."

Other Fair Trade Organizations

Other national NGOs also play a major promotional role, for example, in the United States, Red Tomato (related to AgroFair), Coop America, Global Exchange, United Students for Fair Trade, and others in the Fair Trade Federation, EFTA, and IFAT. In the South, coalitional organizations like Foro Emaús in Costa Rica exemplify remarkable collaborations of fair-banana supporters (chap. 12).

Other groups that champion human and labor rights for various countries could become Fair Trade advocates. In the United States, these might include such organizations as the Nicaragua Network, Witness for Peace, the Network in Solidarity with the People of Guatemala (NISGUA), the International Labor Rights Fund, SweatFree Communities, and the Campaign for Labor Rights, to name a few. They all represent potential agency for a third approach to fairer trade. I elaborate theories guiding possible coalescence of these groups in the following chapter.

CHAPTER 5

Going Bananas as a Social Movement

A POTENTIAL BANANA coalition requires common understanding about development but also demands human interaction. In fact, common understanding and action reinforce each other. Social theorists tell us three components are particularly important: for groups to ally in a social movement they need a "structural opportunity" to come together; this opportunity often requires an activist "network" of people that connects them; and they must realize a common "identity" to sustain themselves. These concepts are summarized in table 5.1.[1]

Opportunities refer to a special convergence of events within social and economic structures that liberate or propel collective action. *Networks*, with emphasis on networking, involve the dynamic interactions between movement leaders, organizations, and opportunities for action. *Collective identity*, as promulgated by writers on New Social Movement (NSM) theory, stresses the subjective aspects of group involvement—that is, how individual participants see and define themselves, and how shared meanings emerge in movements.[2] In recent decades, social theorists have integrated these concepts.

Structural Opportunity

Discovering a "structural opportunity" first requires that participants understand the import of structure in society. *Structure* refers to durable social realities just as it does to visible objects in the physical world, like buildings. Structural opportunities can emerge when tangible forces exacerbate inequalities between people and nations, such as when power alignments, economic crises, or technological changes help reveal elite divisions, interests, and benefits. Theorists describe these occurrences as "political opportunity structures" (McAdam 1982; Tarrow 1989; Smith and Johnston 2002). Once divisions are exposed, the social basis for collective action stems from a contention for power between those who hold it in society and challengers who are excluded (Tilly 1978). Alternative political visions are promulgated that emphasize differences; on the one hand,

TABLE 5.1. Definitions for a Social Alliance

Concept	Definition
Social movement	A group that acts with some continuity to promote or resist change in society
Alliance	A temporary coalition of social-movement organizations that jointly pursue common goals for social transformation
Structural opportunity	Convergence of events that enables change in social relationships and institutions
Network	A set of links between individuals or social units such as groups or organizations
Collective identity	An individual's cognitive, moral, and emotional connection with a broader community or institution

Source: Drawn from Anderson and Taylor 2006; Polletta and Jasper 2001.

for example, poorly paid workers may make demands on wealthy factory owners or disenfranchised political elites claiming state control. On the other hand, corporate managers may threaten employees or small farmers with a withdrawal of livelihoods. Clashes between involved groups encourage both sides to seek political allies, altering the alignments of power and creating instability.

For those potentially seeking a banana "opportunity," power alignments can change with a notable alteration in the methods by which the fruit is grown, purchased, shipped, or marketed. When TNCs that dominate cultivation and export modify banana prices, shipment practices, and distribution patterns, they stimulate reaction. Any change could create fresh opportunities for worker interactions. Corporate lobbying or bribery could also affect the legal frameworks that govern how bananas are taxed and traded.

In evaluating whether potential banana partners should pursue structural opportunities, we have to determine if they consider the contending social forces involved. Do they take into account class relations within the sector and the organizing capacities and strategies of potential alliance members? Do they build on the class factions identified in their analyses and locate pressure points? Do activists like Ramón Barrantes and Anton Bowman consider the class structure of banana production and marketing? For example, the system traditionally dominated by TNCs like Chiquita, Del Monte, and Dole now favors a more diffuse distribution of power that includes national growers and supermarkets, as I have elaborated upon in chapter 2 and will further elaborate upon in chapter 10. Do alliance supporters consider which of these suppliers are more likely to enhance fairer market conditions in Europe, Japan, and North America?

On the supply side, do allies appraise target opportunities by investigating who controls a certain plantation and other plantations or markets linked with it? Do supporters help determine how susceptible those markets are to the various kinds of publicity and promotional strategies I consider in chapter 11? Do activists utilize such information to anticipate repression of farmers or workers who attempt to organize a cooperative or union, or gain a negotiated agreement? Do they adequately pressure TNCs searching structural opportunities in changing markets, nationally endorsed promotional plans, government functionaries, and consumer purchasers to consider activist proposals as positive opportunities for fairness?

Looking at the structure of banana distribution could also help Fair Traders resolve their dealings with banana producers of various size.³ In 2005–6, for example, U.S. demand grew for more Fair Trade bananas. Yet sales dropped. Wild Oats stores and co-op retailers could not procure sufficient quantities to meet demand. In fact, small farmers and producer co-ops could not deliver large shipments of quality bananas to any coordinated market on a regular basis.⁴ Their boats are not refrigerated. Shipping to northern markets could take thirty to forty days, yet bananas ripen in less than a fortnight. By contrast, TNCs like Chiquita and Dole could deliver temperature-controlled bananas in twelve days. Structurally, TransFair, Oké, and AgroFair could have negotiated with the TNCs to ship Fair Trade bananas, while still retaining independence. However, this would have required cooperation, as well as changes in network patterns and significant shifts in identity.

Thus, structural opportunities primarily emerge as cracks in the system of power, but their assessment must also include the strength of resistance from below. For a banana alliance, this involves an evaluation of the various components that Auria Vargas, Anton Bowman, and Jonathan Rosenthal represent, that is, unions, smallholders, and NGO activists. Each group contains internal divisions and interests. An understanding of and respect for differences is essential if a banana alliance is to move together. Small farmers have advocacy organizations, such as WINFA, in the eastern Caribbean, which itself contains organizations from each of the Windward Islands that could cooperate with or challenge WINFA policy. Unions in producing countries are often affiliated with a regional banana-coordinating body, such as COLSIBA in Latin America, which in turn has relations with the IUF in Geneva.[5] A plethora of NGOs support alternative trade and ethical consumer choices, but in different ways—all of which the assessment of structural opportunity must take into account.

Networks

A successful effort at group interaction also involves *networking*. This means stressing the informal links between participants. The network/networking emphasis adds vigor to structural considerations and helps us understand why similar structural conditions might produce differing results (see McCarthy and Zald 1973). The network focus examines how members of one social-movement organization interact with members of another social-movement organization. If interaction occurs on a regular basis, it would verge on structure. If it is irregular but predictable, issues and interests, such as what constitutes fairness, come to the fore. As interaction occurs, a level of trust is established that dynamically generates a movement alliance. Network activists from several organizations then bridge or link, taking the opportunity to challenge organizational barriers (Smith and Johnston 2002).

Movements also form networks and networking because they require resources.[6] Networking provides a communication system, a place to meet, and sufficient materials and personnel to publicize and act on the movement's vision. As Pierre Bourdieu (1990) suggests, social and cultural capital as well as economic capital (a structural element) are essential for a successful movement, and these kinds of capital are often acquired through networking. Different capital combinations lead to various group dynamics and leadership qualities. These, in turn, result in distinct strategies for defining understandings and building coalitions that can either enhance or harm unity among fair-trade promoters, small farmers, and unions.

Usually, a networking focus reveals how contradictory aspects of structural opportunities may be bridged or overcome. In addition to regular channels, informal linkages can offer tools with which to confront organizational barriers. While social-movement organizations may be structurally prevented from taking action because of legal requirements or sociocultural boundaries, through networking they may be able to informally exploit a contradiction within a system of power, such as a discrepancy between a corporation's image and its practice.

By focusing on linkages and cross-cutting ties between groups, for example, between local and international NGOs, network analysis has generated intriguing appraisals of organizational action (see Narayan 1999; Wellman 1999). While mobilizing networks may arise separately among small farmers, union locals, and consumer groups, studies show they are equally likely to emerge between local and international levels.[7] Scholars examining this local-international bridge have described "transnational activist networks" that employ the boomerang model described by Keck and Sikkink (1998): a frustrated local movement organization utilizes the resources of an international network to strike down a barrier at home. Transnational activist networks may still exhibit the inequalities reflective of North-South relations, as Smith and Wiest (2005) have shown. Subordinate networks may resist northern influence via blocking or backdoor moves (Hertel and Minkler 2007). Nevertheless, as Armbruster (2005) has demonstrated in the apparel sector, networks can also operate across borders effectively.[8]

COLSIBA already behaves as a network more than a structure. It functions more like the United Nations than like a TNC: its member organizations hold most of the power, not itself. Yet COLSIBA has utilized its network resources to gain contracts and protect affiliates. In 2006, for example, it employed boomerang techniques to win the reopening of a Honduran plantation that Chiquita had closed because of damage suffered from Hurricane Gamma. Through networking, the IUF has also established worker councils in many locations. Its campaigns have brought notable public recognition and "global framework agreements" with such firms as Chiquita (see chap. 10), Nestlé, and Unilever.

EUROBAN linked northern supporters with southern organizers of antiunion banana plantations in Ecuador. USLEAP has networked with several thousand U.S. supporters and collaborated with EUROBAN and

STITCH to advocate for banana-worker priorities. Banana Link and BanaFair have likewise formed transnational activist networks. Affiliates of smallholder groups like the one Anton Bowman heads, union formations that Auria Vargas and Ramón Barrantes are involved with, and promoters of alternative trade such as Oké Bananas can also be viewed as networks with mobilizing capacity. I will assess whether unions like Auria's and Ramón's engage their local and national federations and confederations; and in turn, whether national federations prompt international federation activity.[9]

Informal network linkages among farmers and consumers can become a resource for material aid and political pressure. In its acknowledged failure to "adequately engage" NGO advocacy groups in launching Fair Trade bananas in the United States (TransFair 2006), TransFair bypassed network endorsement that could have accelerated sales at a crucial juncture. Another challenge, therefore, is whether TransFair will "keep going it alone" (Raynolds 2007, 78).

An emphasis on networking may help fair-trade advocates move cooperatives and labor groups from a parochial to a global focus, thereby gaining the coordinating ability of international federations. A domestic environmental or labor group stonewalled by a local government or company over a company's use of a damaging pesticide or over its abrupt termination of employees could form a transnational activist network with international organizations for further boomerang assistance. Global pressure from abroad may then bring a domestic remedy. Activist NGOs could turn out substantial numbers of network supporters to leaflet stores, picket consulates, and lobby officials, as they have done in the past.

Environmentalists and unions could also tap into networks that exist in communities where they operate. People there often share earlier bonds that encourage solidaristic behavior (Fireman and Gamson 1979). International organizers have discovered fertile ground for labor action in neighborhoods that hold historical working-class connections. When the Foro Emaús church/labor coalition emerged in Costa Rica to resist company exploitation of area workers, lands, and waterways, it depended on community networks created by both groups separately. Although some church leaders espoused a conservative orientation, when corporate layoffs, land displacement, and environmental damage threatened to reduce church memberships, many underwent a conversion toward labor concerns (Emáus 1998a; 2003). Likewise, in the late 1990s *maquila* (assembly factory) organizers in Honduras quickly realized that many of the activist women came from the families of banana workers that had traditionally held strong pro-union sentiments (interview with Fieldman 2002; Pearson 1987).

Identity

The third consideration for achieving common understandings and action is the way alliance members form at least a partial and temporary *identity*. Various participants may understand it somewhat differently; and in fact, disparate understandings may assist group unity (Tsing 2004). Yet self-perceptions must be sufficiently common that people can say, "*we* are working together." Clarifying this process makes the analysis of identity the most challenging among the three requirements for alliance building.

New Social Movement theorists stress an interactional and "intersubjective" process of identity formation (Habermas 1984, 1989). This means it arises through social exchange and negotiation. Jürgen Habermas and others describe movements in the latter twentieth century as often led by people who did not hold important government and economic positions. They acted in "public space," that is, in streets, parks, and locations that ordinary people inhabit.[10] New movement leaders created unity by emphasizing social or cultural values such as peace or ecological restoration, but also religious fundamentalism. As part of identity "negotiation," they urged followers to participate in overt social behavior that demonstrated their self-definitions. Critics claimed that such public actions were often more symbolic than practical, more an idealization of common goals than an achievement of durable organization. However, manifestations of religious belief in the twenty-first century have bolstered identity analysis.

Those theorizing about identity believe that "intersubjectivity," or shared human experience or insight among group members increases the dialectical interplay between understanding and action. Social scientists have shown that as collective agency increases, participants broaden membership awareness (Rose 2000). U.S., Canadian, and European activists discovered a congruent application when third world workers warmly responded to symbolic support and grew to appreciate global corporate campaigns. One example was a welcoming reception striking maquila workers in Puebla, Mexico, offered a delegation from the United Students Against Sweatshops (USAS) (interview with Traub Warner 2001). Similarly, antiglobalization demonstrators have tapped symbolic expressions and subjective feelings to challenge objective inequalities in Latin and other countries (Escobar and Alvarez 1992; Alvarez, Dagnino, and Escobar 1998; Klein 2000; Barlow and Clark 2001; Featherstone 2002; Wickham-Crowley and Eckstein 2003; and others.)

Identity-related movements devote special attention to a key component for alliance building—how the participants subjectively understand the message. Such an understanding encompasses nonrational aspects, feelings, and beliefs. Goffman and other theorists have characterized this process of identity formation as "framing" (Goffman 1997; see also Snow and Benford 1988; Snow and McAdam 2000).[11] The framing emphasis cues in on how leaders from various locations might build identity to forge a common organizing strategy, but also on what being an alliance member *means* for participants. Both Melucci (1996) and Goffman (1997) have demonstrated how frames are not simply imposed rationally. Instead, they emerge from an interactive process between prospective participants that defines network culture and fields of action.

Thus, such framing can be of two sorts. In the first sort, opted for by movement spokespersons, identity is strategically applied to gain and nourish members in a common project (Snow, Rochford, Worden, and Benford 1986; Snow and Benford 1988; Tarrow 1998). Leaders make a convincing argument about an unjust situation, and why movement members are the ones that can change it.[12] The second sort of framing, however, is not seen as a strategic or instrumental imposition but as a self-defined exercise that emerges from the ground up of what people believe or value. These frames can be sustained by "identity rituals" like sit-ins and demonstrations (Polletta and Jasper 2001). Some describe this type of framing as an individual's own take on a situation (Snow and Benford 1992). It can be about a shared status, a sense of justice that is imagined or experienced, or a desire to be perceived as loyal and committed. This type of framing implies that movement participants forge new identities that are flexible, inclusive, and less rigidly defined than the first kind of framing (della Porta and Tarrow 2005). Whether frames are guided or spontaneous, though, the who-we-ares of one social-movement organization can easily conflict with those of another, with strategic implications.

Testing a Fair-Banana Identity

The identity focus has directed activist interest to bananas, for example, to the internal conflicts of WINFA farmers who struggle to become both Fair Traders and union allies. As they experience declining markets, some individual island farmers and associations have viewed WINFA as another middleman that siphons off Fair Trade premiums. Yet just as the Fair Trade identity of small coffee producers evolved with support from NGOs like the Havelaar Foundation, so it takes time for leaders like Anton Bowman and Nioka Abbott to assist WINFA farmers as they reframe an "identity shift" for producing bananas that includes a self-resonance as members of an integrated association.

Likewise, the identity approach can elucidate how producers like George DeFreitas, Jiménez Guerra, and Auria Vargas rally those around them to collective action.[13] This approach helps clarify COLSIBA's behavior as it brokers a plethora of reactions to protect affiliate unity (its Colombian unions have proved to be a special challenge). Via an identity emphasis, groups represented by people like Ramón Barrantes and Jonathan Rosenthal have also built coalitional awareness as a force that helps sustain their social-movement organizations. As partial identities emerge around Fair Bananas, a similar force could aid participants in becoming a "social movement coalition." An identity focus could help unite consumers and producers when CLAC reminds them that feeding one's own people must take priority over any export project. It could guide responses to COLSIBA, whose representatives played an early role in discussions around Fair Trade, urging advocates that true sustainability meant respecting both small farmers and workers in all countries.[14] Coalitional identity concerns the IUF, another early voice that emphasized rights as a crucial Fair Trade benchmark. The International worries that distributors of Fair Trade products may be oblivious to union freedom. These are key identity issues that, left unresolved, could sabotage any Fair Trade alliance.

Certain EUROBAN members such as France's General Confederation of Labor (Confédération Générale de Travail [CGT]) aided FLO proponents in including rights in their definition of Fair Trade; however, other EUROBAN members retain their own discrete interpretations about what freedom of association entails. With its sensitivity to a wide customer network, the Oxfam consortium represents a potential alliance component that has long articulated a Fair Trade identity. Oxfam Wereldwinkels (Belgium) was a FLO founder, and Oxfam affiliates within individual countries have become leading FLO advocates. Nevertheless, the Oxfam partners have not yet fully resolved their position on union bananas. Some espouse creating a combined council or assembly of employees and managers, that is, a "joint body" as equivalent to free association, a position anathema to unions. Others question how carefully FLO and other FT affiliates assure fairness among their suppliers. On the other hand some Fair Trade certifiers and suppliers vehemently oppose COLSIBA's definition of labor rights as equivalent to union rights.

Thus, identity issues remain as a daunting coalitional challenge. In the United States, TransFair's promotion of Fair Trade to corporations as a way of reducing activist pressure reveals an insensitivity to identity concerns. TransFair's "more volume, more farmers served" approach likewise indicates a naïve disregard for Fair Trade's historical self-definition in which small retailers and producers have viewed the program as a genuine alternative to mainstream marketing (Raynolds 2007, 75–76). If a coalition should decide to "break into the mainstream supermarkets" with "the participation" of major firms as TransFair (2006) suggests, a significant identity transformation would be required. NGO involvement in *monitoring* company codes of conduct raises further identity difficulties for any alliance (see chap. 11). USLEAP stresses that responses to labor violations must be determined by the local union involved, a principle that at times has kept USLEAP from promoting an international campaign because workers on the ground have not requested one.

Thus, for a banana alliance to be fully reframed—extended or amplified—to incorporate shared "communalities," several phases are required:

1. Banana workers must demonstrate a broad identity in regard to an international fair-trade strategy that encompasses and promotes small-scale production

2. Small farmers need to achieve a fair-trade orientation that involves worker rights.
3. Producers, certifiers, distributors, and retailers need to establish a consultative process for evaluating appropriate scale.
4. Fair-trade promoters have to clearly understand and commit to union principles at local, distributional, and international levels.
5. All partners must vigilantly oppose export agriculture from interfering in local food production, indigenous land priorities, and gender equity.

Given the various interests and perspectives involved, achieving these stages toward a coalitional identity is no easy chore.

Joining Development, Structural Opportunity, Networks, and Identity

In recent years, social-movement scholars have labored to integrate structural opportunity, networking, and collective identity into an explanatory framework. David Meyer, Nancy Whittier, and Belinda Robnett (2002) present cases that link opportunity and identity. Patricia Hipsher (1997), Laura MacDonald (2005), Francesca Polletta and James Jasper (2001), Jackie Smith and Dawn Wiest (2005), and others find that identity is intimately associated with networks poised to respond to opportunities.[15] Following the methodology outlined in chapter 1, I use an integration of these concepts to evaluate potential farmer-worker-consumer understandings and interaction in the banana sector. It is unlikely that any alliance would include advocates for small farmers who manifestly reject links to plantation production, or environmentalists directly related to corporations. Since such individuals or groups could either encourage or discourage the formation of such an alliance, I take their views into account. But my test for potential alliance members is those that promote a view of sustainable development that incorporates a strong sense of agency from the grassroots.

Although each of the social-movement organizations related to trade fairness is committed to sustainable development, and each demonstrates elements of structural opportunity, networking, and fair-trade identity, their unity will occur only with attentive convergence. While there exists some common understanding of peasant exploitation and the benefits of independent worker organizations, convergence requires additional steps to pinpoint salient opportunities for exerting pressure and resisting the brutal repression that corporate power and institutional forces can unleash to damage any competing alliance. Only a unified campaign can endure.

Summary

In chapters 4 and 5 I laid out potential actors and theoretical components for a convergence around fair bananas. Development theory that elucidates the role of agency from below can guide proponents, but a unifying movement also requires a target opportunity, a mobilizing network, and a negotiated identity. Besides exposing cracks in the banana system, a structural opportunity could involve on-the-ground organizing led by local leaders. An alliance would more likely occur if its leaders could mobilize a pre-existing network, first, within the levels of union or smallholder affiliation, second, among family and community-related linkages, and third, with international unions like the IUF, peasant support groups, and NGOs acting as movement organizations. Networking could then empower local workers and summon additional resources.

Moreover, a coalition needs to solidify a membership identity by aiding participants who subjectively negotiate their movement involvement. Even if one assumes producers and workers have similar structural opportunities, both groups still wonder whether it is in their interest to combine approaches to improve life conditions; an identity focus helps resolve their responses. NGOs and participating retailers are also asking themselves just who should be part of Fair Trade, apprehensive that large TNC involvement could jeopardize its principles. Some FLO participants suspect that unions might usurp funds for patronage purposes. However, many argue that distributing Fair Trade goods to a wider public will convey more benefits to small farmers and workers. They urge consumers and students to develop a Fair Trade consciousness as a moral project to increase commercial outlets willing to buy from registered producers on their Fair Trade terms.

In many areas, the promised equalization of free trade and investment has not brought benefits. Environmental conditions have improved for some but not for others. The huge discrepancy between northern incomes (averaging above $20,000 a year) and the incomes of southern workers and farmers (averaging below $500 a year) has largely persisted. Fair banana proponents like Nioka Abbott, Ramón Barrantes, and Jonathan Rosenthal may agree that change could occur if those who bought bananas added social and environmental considerations to their choice. That they remain skeptical about whether they can build on this opportunity to network and claim a partial identity is exactly the challenge we continue to pursue.

CHAPTER 6

The Persistent Banana Environment

REBELLIONS BY THE earth and its waters have precipitated a burgeoning environmental consciousness. Daunting corporate abuse has stimulated resistance—most obviously from nature but also from workers and consumers. Their mobilizations to defend the global habitat first forced banana TNCs to search for disease-resistant seedlings and later, to promote ecologically friendly production as a marketing strategy. In this chapter I discuss how large-scale banana cultivation easily led companies to exploit nature, but also how their maneuvers offered a remarkable structural opportunity that impelled farmers, unions, and even some corporate managers toward notable environmental improvements. When the Fair Trade network incorporated strong ecological protections with socioeconomic guarantees, it further bolstered the movement toward healthily grown bananas.

The Environmental Impact of Commercial Bananas

In the first half of the twentieth century, banana expansion went hand in hand with environmental change as well as with the political and economic transformations described in chapter 2. Companies took over large swaths of terrain, forced out local farmers who were growing food, cut down forest habitats, and destroyed wildlife. Although growers arranged production to fit the banana-husbandry cycle, their rules transformed the naturally occurring plant into an artificially spaced crop.

Laborers prepared the earth for small seedlings that they fertilized and weeded until the stems sprouted mature stalks that produced about 150 banana fingers. Workers made sure the fruit had nutrient soils, reduced weeds, a certain amount of water, and protection from invading pests. After about a year and a half, at just the right moment prior to ripeness, the producers then harvested the stems (see fig. 6.1). They either boxed the fruit on the spot or delivered it to a plant where packers would clean and separate the finger hands (bunches) into clusters, apply

FIGURE 6.1. Del Monte banana harvest, Guatemala. (Photo by author)

fungicides, and place the finger hands in forty-pound boxes for shipment abroad.[1] Remaining field stalks would then be cleared to make way for "daughter" shoots, thereby repeating the process.

Bananas increasingly became a monocultural crop that transformed landscapes and supplanted diverse plant species as companies continually imposed stricter requirements for cultivation, harvest, and delivery. Companies redesigned waterways for irrigation and drainage in a manner that increased soil erosion. As the fruit became more susceptible to various pests such as insects, fungi, and competing weeds, the TNCs broadsided the fruit with pesticides, severely damaging the ecological system and harming workers and communities. Nevertheless, these environmental practices, summarized in table 6.1, created impacts which became structural opportunities that stimulated fresh networks and group identities.

In the late 1920s, a disease hit Central America's Atlantic coast that virtually wiped out United and Standard Fruit's plantings of Gros Michel bananas, the variety most companies shipped abroad. Soon, the TNCs

TABLE 6.1. Environmental Impacts of Monocultural Banana Production

Practice	Impact
Land clearance	Deforestation, loss of species diversity
Irrigation	Erosion, waterway contamination
Pesticide applications	Water, resident contamination; fish kills
Stem casings	Plastic wastes
Packing and shipping	Additional toxics; worker endangerment

abandoned Atlantic coast production and began afresh with new clearings along the Pacific, and a new banana strain, the Cavendish. However, the Pacific expansion attacked ecosystems in a way that did not allow soil recuperation. Deforestation exacerbated flooding; pesticides poisoned lands, and irrigation intensified erosion, sedimentation, and water shortages that further harmed arboreal vegetation (Harari 2005a).

In the mid-1980s, the big three TNCs, Chiquita, Dole, and Del Monte, inaugurated a third wave of banana expansion to take advantage of anticipated markets in eastern Europe and China. Returning to the Atlantic coast, the companies garnered local government support by trumpeting the ideology of "beneficial progress." Enamored Costa Rican officials issued a "Plan for Expansion of Banana Lands," which the National Banana Corporation utilized to confiscate an additional eighty thousand acres as companies cleared tropical rain forests (May 1998). Similarly, in coastal Ecuador, companies extended their plantings from hills to seaside, removing forests and polluting waterways in between. This time, throughout the region, peasants and environmentalists loudly complained about lost trees, soils, food, and nutrients. Workers warned about the expansion's effect on community health (Holl, Daily, and Ehrlich 1995; Hermosilla 1998, 8; Harari 2005a).

Pesticide Applications

To avoid blights like the Panama disease of the 1920s, scientists sponsored by United Fruit and other TNCs experimented with new banana strains[2] and tested various chemicals to prevent pest invasions.

In early decades, the most widely applied pesticide was copper sulfate, a blue-greenish liquid that companies sprayed on the plants to prevent black sigatoka, a fungus that attacked plant leaves and caused bananas to ripen prematurely. It especially affected the Gros Michel variety. In 1935, United Fruit installed an extensive pipe labyrinth and began weekly sprayings of its "Bordeau" mix of copper sulfate, water, and lime. Soon the company had built tower systems to pump the spray aloft. The spume turned the plants, workers, and surrounding area a persistent light blue. Workers quickly became aware of copper sulfate's detrimental effects and petitioned United Fruit to take precautions; however, most unions did not officially endorse their solicitations, fearing the company would withdraw jobs (see Marquardt 2002; and chap. 7). Yet the sulfate residue did not decompose. By the 1940s, TNCs had realized that it could damage shipping vessels, so they carefully cleaned the bananas in muriatic acid, then washed them in water before packing. However, it took the agrifirms another ten years before they gave any thought to adopting similar precautions for the workers who handled the fruit. At that point they had fully replaced the more disease-prone Gros Michel with the Cavendish banana, yet this would give them only a temporary respite (Koepple 2007).

So then, in the late 1960s, the TNCs introduced a new round of *toxic "cocktails"* that affected each aspect of plant growth: herbicides to knock out competing weeds; then nematicides to attack nematodes—tiny wormlike creatures that feast on roots within the soil; and finally, fungicides to eliminate plant blight. At the packing plant, the companies washed the cut bananas in chlorine baths to remove latex, field chemicals, and manganese and iron, which caused spotting. They painted more fungicides on the stems to retard crown rot as they prepared the fruit for shipment. The entire process involved such known poisons as Counter and paraquat to handle weeds, and DBCP,[3] which Shell and Dow Chemical formulated to kill nematodes. Between 1968 and 1979, companies applied more than 1.3 million gallons of DBCP to banana plants in Costa Rica alone. The companies also encased the growing fruit with plastic bags saturated with insecticides. They aerial-sprayed banana plants to attack the persistent black sigatoka fungus.[4] When injected in the soil, DBCP improved banana yields. However, the aggressive use of nematicides and fungicides caused a decline in amphibian and aquatic populations near banana sites.[5] DBCP affected human populations that drank or washed in the water. Residues appeared in houses, schools, and recreation areas. Finally, in 1979, the United States banned DBCP for domestic use.

Owing to TNC influence, Central American governments resisted similar measures, making it difficult to document DBCP field-effects. Nevertheless, men and women reported increased sterility. By the late 1980s, the major companies faced substantial lawsuits (see Murray 1994). In 1997 Dow paid $22 million in an out-of-court settlement with twenty-four thousand workers who claimed DCBP had caused their sterility. After further demonstrations in 1998, the Costa Rican government investigated uncompensated DBCP impacts and urged an indemnity fund, additional medical attention, and pension relief (Taylor and Scharlin 2004, 118). In 2005, COLSIBA's outspoken union leader, Iris Munguía, railed against the "irreversible health damage" documented by "more than 15,000 lawsuits against TNCs and producers and their use of DBCP."[6] In 2007, a Nicaraguan court and a Los Angeles jury ruled that Dole was liable. The latter assessed $4.9 million for damaging workers, a charge the company forcefully resisted (chap. 13).

For many years, company managers paid little heed to the impact these toxic materials had on the environment or the harm they might do to workers. Plantation supervisors gave minimal instruction about how pesticides should be properly mixed, applied, and stored and rarely offered protective equipment to keep employees safe. Yet the spraying often nauseated and ultimately sterilized field hands. Packing-plant chemicals caused women to develop headaches, respiratory problems, skin lesions, and nail fungi. Some developed traceable cancers, miscarriages, and other gynecological difficulties (see Wesseling et al. 1996; Astorga 1998; Center for Environmental Studies 1996; Foro Emaús 1998c and 1998d; Hernandez and Witter 1996; Hirsch and Aguilar 1996; Lamb 2008, 11–12; Mortensen et al. 1998; Wheat 1996).

As chemicals polluted air and waterways, community groups articulated grievances that increasingly gained public attention. In Costa Rica, for example, workers, small farmers, and religious groups rallied together as Foro Emaús, to demand redress for banana areas and residents. By publicizing DBCP sterilizations, corporate contamination of Atlantic waterways, and TNC deforestation, Foro publications, forums, and demonstrations eventually reached a sympathetic audience in Europe, inspiring environmentalists and others to send delegations.[7] As European consumers became more aware of the ecological havoc wrought by monocultural banana production, the TNCs began to address their worst practices. They deflected criticism by ceasing use of DBCP and paraquat and making other limited changes. As I noted in chapter 2, in 1992 Chiquita began the Better Banana Community Agricultural Program under the guidance of a U.S. environmental NGO, Rainforest Alliance. After an initial prodding, Chiquita's corporate managers envisioned the approach as a marketing opportunity, a Smart Alliance, as environmental consultants J. Gary Taylor, and Patricia Scharlin, who served on the Rainforest board, entitle their engaging study. Yet Chiquita continued applying a host of toxic chemicals.[8] Eventually, the Rainforest Alliance implemented its Sustainable Agricultural Network (SAN) with a comprehensive checklist to certify environmental practices. Convinced by earning 20–30 percent price premiums on its "quality bananas" in Europe, Chiquita committed to having all its fincas certified (chap. 11). Dole followed with guidelines based on a new International Standards for Organization environmental-management-appraisal system (ISO 14001) and made recognizable environmental improvements in Costa Rica.[9] As U.S. consumers took interest in banana conditions, Del Monte also adopted ISO 14001 (Masibay 2000). In Costa Rica, fruit company managers created "an Environmental Banana Commission and promoted awards to the packing plants that reforested the riverbanks, collected plastic bags and offered proposals for recycling toxic waste" (Palencia 1997, 28). By 1997, the big three TNCs had implemented certain safeguards for applying fungicides and handling pesticide-laden plastic bags. They installed governors to limit the drift of aerial spraying and sharpened rules for treating liquid and solid wastes.

But TNC scrutiny accelerated. The ISO 14001 system did not contain performance-based measures, much less an impact assessment of worker health (see Krut and Gleckman's critique [1998]).[10] Local environmental-community coalitions like Foro Emaús remained suspicious that the agrifirms were promoting environmental-friendly labels to improve their image "without having substantially altered their use of pesticides" (Palencia 1997, 28; see also Vargas 1998, 7). European unions and consumer groups feared that companies would invoke "environmental certifications" to bypass stringent labor requirements. EUROBAN banana activists suspected that the Rainforest Alliance's Better Banana/SAN certification made promises Chiquita could not fulfill, especially regarding employee treatment (Murray and Raynolds 2000, 70). Although the

TNCs had substantially reduced the most toxic chemicals, they paid less attention to chronic exposures.[11] In Costa Rica, companies still applied more than fifty spraying cycles a season, twice the number used in Ecuador. Ostensibly, managers shifted workers out of fields slated for spraying for thirty-six hours,[12] but despite new nozzles and controls, the aerial spumes drifted. Pilots admitted honoring flight schedules no matter what way the wind was blowing (Harari 2005a). Some studies showed only 10 percent of the spray actually reaching the banana leaves.[13] Citing deadly chemicals, worker poisonings, and fish kills, Foro Emaús organized press conferences, street demonstrations, and research verifying that banana packers suffered twice the genetic damage of others (Escofet 1998; see also Wesseling 1997; Naranjo 1999). "Almost all of us were poisoned," one woman packer explained to the Foro. "I was affected at least three times, causing me stomach pains and vomiting" (*La Voz de la Mantí* 1999).

Companies Further Address Environmental Concerns

Eventually, consumer criticisms of corporate "greenwashing" brought more serious company efforts. Chiquita further improved environmental practices,[14] cutting pesticide use by half. Although it would not refrain from aerial spraying against sigatoka, it promised to apply only pesticides approved by the U.S. Environmental Protection Agency (EPA).[15] The agrifirm inaugurated water testing, plant filtration, and recycling. It stepped up the reprocessing of discarded (culled) bananas into pulp, the composting of organic waste, and the creation of landfills to cover nonorganic materials. It enhanced native plantings between fields and areas of human settlement. It trained employees in Integrated Crop Management (ICM), which emphasized a much more restrictive and safe agrichemical transport, storage, and application. Even if worker and community health were not yet fully safeguarded, many employees and community leaders believed that Rainforest certification had reduced Chiquita's use of Counter, paraquat, and certain other toxics. The company placed ornamentals around packing sheds to make them more aesthetically pleasing. It also offered more extensive protective equipment to workers, although they did not always utilize it.[16] The company installed laundries so that contaminated clothing could be washed immediately.

In Sixaola, on the border between Costa Rica and Panama, union leader Auria Vargas Castañeda felt that, owing to Rainforest certification, company aircraft had become

> more careful about spraying over our homes. When they fumigate, there are warning signs. For example, supervisors now publicize that they will spray one field in the morning while the workers are in another field, and then shift in the afternoon. In packing, they use spray boxes to contain fungicides and pregnant women are not assigned to work with the boxes or stamp seals on the fruit. At least when officials are present, managers have to abide by the rule that we must go outside to drink water. Workers come to me seeking a change in jobs [for environmental reasons], which I can now pursue. So the certifications have brought one good thing—a higher level of mutual respect. The managers used to yell at us a lot. Now [she laughs], I yell at them. (interview with Vargas Casteñeda 2004)

Such TNC monitoring programs, along with NGO pressure from groups like Foro Emaús, inspired other national producers and promotion bodies to take up the mantle of environmental protection. In 2001, Costa Rica's private banana-promotion agency, the Corporación Bananera Nacional (CORBANA), publicized "Rules of Conduct for the Prevention of Labor Risks."[17] However, CORBANA offered no benchmarks or data that demonstrated that their rules were followed. Likewise, in 2003 Ecuador required companies to submit approved health and safety policies, but only eight of the largest two hundred firms complied (Harari 2005a). Chiquita and Dole also extended their certification program to the plantations owned by their independent producers (see chap. 11), but Rainforest subsequently discovered higher chemical residues there (Whelen 2003). Subcontractors were not careful about protecting field hands, passing the risk on to them, and TNCs still incurred accidents. In 2003, Dole spilled two thousand liters of Bravo 72 in Costa Rica (Harari 2005a, 45), and area cancer rates remained high (Wesseling et al. 2002).

Nevertheless, by mid-decade many banana companies had made notable changes. COLSIBA representative Gilberth Bermúdez acknowledged that the TNCs had "brought improvements with at least 6 different programs—Better Banana, SA8000, ISO 14001, company codes, U.S. regulations, Eurepgap" (interview 2004). Geographer Dr. Carrie

McCracken agreed that in Costa Rica, "even the independents are adopting a lot of safety measures. Six months ago I did not see women wearing gloves in the packing sheds, but now most do. Managers have stopped using the dirty dozen pesticides, and made buffer zones for aerial spraying between worker housing and roadways. Fincas are recycling plastic bags [used to cover the fruit], although they still accumulate" (interview with McCracken 2004).

Still, commercial monocultural banana production could gloss its environmental record only to a certain point. When climactic disasters occurred, cleared banana lands exacerbated flooding.[18] Companies still sprayed to control nematodes and black sigatoka.[19] Foro communications director Hernan Hermosilla insisted that "while there is more careful registration and control over chemicals, there is no decrease in amounts. After five years, pesticides still represent 30 percent of per box production cost and fish kills persist" (interview with Hermosilla 2004).[20] In September 2006, residents and tourists near the Estero Negro River in Limón Province once again cited a huge fish kill that generated an intolerable stench and rat infestation. The Costa Rican minister of environment and energy acknowledged it as a regular occurrence, likely from banana chemical contamination.[21]

"The Integrated Pest Management approach tells companies what cocktail to use for sigatoka, but it is constantly changing because resistant strains keep developing, and spraying still occurs twice a week. You can find fumigation residues on homes, especially in Sixaola [where Jimenéz Gomez and Auria Vargas work]." McCracken watched workers coming off fields after spraying had begun, and one Dole manager told her that 'zero people' in the fields meant 'almost no people' (interview with McCracken 2004).

Environment and Fair Trade

Thus, despite TNC improvements, ecological damage wrought by monocultural banana production remained a structural opportunity. Fair Trade and organic growers stepped forward with alternatives. The Fair Trade approach combined stringent environmental guidelines with its social criteria. In the mid-1990s, ecologically aware consumers caused the FT movement to incorporate rigidly defined nature and health protections under two rubrics: "current requirements," and "progress goals," the latter allowing a grace period of two years for improvement.

Environmental Development Standards

Small producers like George DeFreitas wrestle with FLO environmental standards, but so do Gariba Musah and the unionized workers he leads at Volta River Estates in Ghana, which produces Fair Trade bananas with government support. For approval, both farmers and workers must be trained in FLO's environmental regulations. While humans everywhere attempt to skirt rules, once Fair Trade participants are trained, they are less likely to be caught in the "certification frenzy" that Auria Vargas Casteñeda described in busily preparing to impress Chiquita auditors. As DeFreitas, Musah, Nioka Abbott, and others become involved in ecosystem conservation, erosion control, water protection, and waste recycling, they learn integrated chemical management. They plant other crops to prevent notable pest infestations from spreading, and gradually replace agrochemicals with organic fertilizers and biological disease-control agents.[22] When they find a disturbing problem, they notify a Fair Trade advisor, who may permit careful spraying of certain plants. When root and soil tests demonstrate a specific need, and where nonchemical, biological approaches are shown to be ineffective, FLO offers a two-year plan for reducing nematicide applications. It usually permits only the application of granulated formulations and attempts other steps to cut in half or eliminate other chemical agents.[23] In the integrated-chemical-management model, crop and species rotation can sometimes avoid black sigatoka, the fungus so devastating to production in Central America. FLO recommends manual and, if necessary, helicopter applications to control spray drift. With limited exceptions, FLO plans a phaseout of plastic bags impregnated with pesticides that cover the hanging fruit. When crews handle chemicals directly, they must fill and mix containers on impermeable ground and dispose of container wash via filtered drains. Progress goals give producers two years to construct a filtering plant for wastewater. Through the entire process, FLO participants learn to devote more careful attention to each phase of cultivation, further linking their identity to the land and water around them.

Some FLO growers became fully organic, as exemplified by Heriberto Custodio and his co-workers in the Dominican Republic. Heriberto was selected as president of the Finca 6 Association near Azua, close to the Haitian border.[24] About eight hundred small-scale banana producers populate the region, with two-thirds of them growing FT bananas (Shreck 2002). The 275 households of Finca 6 sell to Savid, S.A., a Dominican "alternative trade" exporter. All around are flourishing banana plants that lived without chemicals of any kind. As Porfirio Acosta Gil, the production manager at Savid, explained "The prime requirement is for meticulous tending of the plants, cutting away dying leaves, keeping the topsoil clean, removing the flowers from the bunches at the right time, covering them with reusable, chemical-free plastic bags to encourage growth and prevent damage from birds and surrounding leaves" (Ransom 1999: 90–91). Christoph Meier, owner of Finca Girasol nearby, also stressed the importance of understanding soils and soil-plant interactions (Boshart 2004).

In earlier times in Ázua, the villagers had been burning the forest to sell charcoal. "The forest is a living thing, and we were killing it. We knew it couldn't last long. Our environment is alive and we want to protect it. But we had no choice, no other way of making a living," explained Angel Custodio, a co-producer with Heriberto (Ransom 2001, 89). In Ázua, the Dominican government assisted by constructing an irrigation project in 1983, and then helped build the entire village and offered land incentives. At first Heriberto, Angel, and the others grew just maize and cassava, but soon they discovered bananas could do well in the limestone-based soil. They learned about Fair Trade from Max Havelaar and Solidaridad representatives in the northern region.

Owing to dryer climate conditions, Caribbean farmers have largely avoided sigatoka since they have less rain to spatter the fungus, but they face a variety of other pests. During the 1980s however, many island growers freely doused their plantings with herbicides, and the runoff contaminated streams and rivers. Grossman (1998, 192ff) demonstrated how St. Vincent associations encouraged the smallest farmers to apply pesticides to meet the marketing requirements for standardized, spotless fruit. Under Fair Trade guidance, independent producers can instead plant crops such as canavalia beans between their banana seedlings. The beans serve as a cover/green manure crop for young banana plants at the

Girasol finca near Heriberto's Finca 6 Association. Calves roam among the plantings, providing fertilizer and weed control (Boshart 2004).

Besides their efforts to do away with agrochemicals, Heriberto, George, and Nioka were tutored in planting methods that minimize erosion of soil into the nearby channels that have been cut to irrigate the banana plants. In turn, these channels feed ditches that drain into rivers that flow toward the sea. Reddish brown soil wash contaminates ditch and river water in many banana areas. To avoid this, FLO adjusts its planting recommendations to various ground inclines,[25] and urges producers to mulch strips of vegetation along the drains.[26] It encourages the Finca 6 practice of members routinely taking turns working on one another's parcels. In addition to creating buffers, they cut back surplus shoots, clean channels, and lop off unhealthy leaves (Ransom 2001, 93).

Finally, since banana production generates considerable waste (some say two tons of waste for every ton produced!), Heriberto, George, and Nioka received instruction on ways to enhance recycling. If the banana clusters are brought to packing sheds and stems are cut off, unripe or unsuitable culls are tossed into a huge waste pile that can become a daunting disposal problem. Like Chiquita, Fair Trade producers have discovered methods to separate out what is suitable for puree or compost. Stems and plastic bags are more problematic. Recycling plants often do not efficiently process all the available bags, and many end up in area waterways. FLO's rule for the disposal of nonrecyclable material is to bury it in an area two hundred meters away from any water.

As FLO certification has expanded, small co-ops have struggled even to meet baseline goals. In 2006, farmer board members succeeded in altering FLO requirements to make their phase-in less onerous and to allow flexibility in applying regulations to local conditions. Despite these modifications, FLO rules offer a much healthier alternative to corporate banana production.

Environmental Practice as Opportunity, Network, and Identity

A healthy banana ecology exemplifies a structural opportunity that can potentially rally small farmers, unionists, and consumers into a joint network. Their environmental understanding potentially represents an element of shared identity. The corporate-imposed monocultural banana system has wrought systematic destruction but also created an opportunity for worker, community, and public reaction. While network formation and identity enhancement takes time, as Shreck (2002) notes, farmers like Heriberto Custodio, Nioka Abbott, and George DeFreitas, have internalized a greater care for the natural world, reviving lessons passed down by their ancestors but largely forgotten when the big agribusiness TNCs invaded the region.²⁷ Unionized workers and supportive groups such as Foro Emaús have likewise expressed this awareness. Following public reaction, TNCs made efforts to improve their environmental behavior. However, Chiquita's sporadic compliance generated mixed reviews, and Dole's environmental reforms neglected social aspects.

Fair Trade, by contrast, had originally emphasized social considerations. Its small-scale, local-control approach also contained environmental principles that it more clearly articulated in the mid-1990s. In pursuing FLO's ecological requirements, small producers and workers enhanced their environmental consciousness. In dialectical interplay, this enhancement motivated further improvements at Chiquita, Dole, Del Monte, and among independent producers. NGO initiatives and consumer awareness as expressed through Fair Trade purchases likewise stimulated TNC reforms.

The success of Fair Trade products has progressed to a point where the companies now want their fruit to be identified as Fair Trade. In 2005, Chiquita spokesman Mike Mitchell claimed that nearly half the bananas the company distributed met "the fair trade guidelines of Social Accountability International" (Dillon 2004) and Chiquita took steps in 2005 to gain official FLO approval in Honduras (see chap. 10). Whole Foods, the 145-store supermarket chain in the eastern United States, collaborated with Earth University in Costa Rica to certify its bananas as environmentally sustainable. In 2003, Dole inaugurated an arrangement with FLO to validate certain European shipments. Spokeswoman Freya Maneki said the company was exploring "the feasibility of further collaboration with the Fair Trade movement" (quoted by Dillon [2004]). Yet, as I will demonstrate in future chapters, small farmers, workers, and activist NGOs did not believe that large-scale banana production had overcome its limits to fully protect the environment. It also faced another structural challenge: the rules of trade.

CHAPTER 7

Conundrums of the Banana Trade

FAIR TRADE CERTIFICATION is a relatively new way to support the people who grow bananas, but attempts at trade fairness have an early history. Small farmers, banana unions, and a number of government and corporate officials have long favored an equitable trade policy. They believe rules that balance production and consumption are crucial, especially for handling perishable commodities efficiently. Trade rules solidify understandings about how much product can be marketed from a particular country, and they help prevent oversupply. As importantly, balanced trade offers a rational method for sustaining economic development in areas historically committed to banana livelihoods. For farmers like Juan Quenteña, George DeFreitas, and Nioka Abbott; and for unionists like Jimenez Guerra, Auria Vargas, and Ramón Barrantes, export-import regulations strengthen marketing and bargaining agreements because they stabilize the conditions within which the agreements are framed.

In addition, free trade can disadvantage farmers and workers. When regulations are loosened or disregarded, product gluts can undercut farmers' and workers' ability to negotiate prices and wages. Product overage may temporarily benefit consumers, but when bananas rot on ships, workers are laid off, small producers lose, and immigration dilemmas accelerate. As a consequence, peasants, unions, and enlightened policymakers show a keen interest in keeping banana trade output and prices within reasonable bounds. Trade regimes offer the opportunities to accomplish this.

Overview of Banana Trade Structures

The United States and Europe are by far the largest importers of bananas (see table 7.1). In 2005, the United States and Canada purchased 29.4 percent of the world's net imports in an "open trade regime" that charged no tariff. The European Union purchased 39.6 percent, via a complex system of quotas and tariff allocations that reflected a rule-based trade policy. I briefly considered the evolution of the European system in chapter 2.

TABLE 7.1. Top Banana Importers, 2005 (000 metric tons)

Country	000 Metric tons
EU	5,843
U.S.	3,917
Japan	1,066
Russia	1,002
China	429
Iran	418
Canada	416
Argentina	296
Korea	253
Saudi Arabia	220
World	14,749

Source: FAOSTAT 2006.

Not only does the United States not have a rule-based system, but beginning in the early 1990s, the United States gradually undermined Europe's fairer-trade orientation by indefatigably demanding an end to trade barriers. Since bananas rank among the four most globally traded foods, the changes advocated by the United States have had major consequences. They influenced the EU's shift from a quota to a tariff-only system in 2006. However, European consumers still pay a required premium for bananas.

Fair Traders have their own trading rules, which were briefly outlined in chapter 3. These include two price guarantees: a FLO-determined price for each country and a special premium for development. Nevertheless, global market prices and trading regimes influence the FLO-determined country price, and FLO-certified bananas must compete with other bananas within this framework. For example, when excess bananas hit the world market in December 1999, Heriberto Custodio and his co-workers in the Dominican Republic experienced a 33 percent drop in the Fair Trade price (Shreck 2002). An understanding of trade issues is therefore an essential requirement for any fairness alliance. In this chapter I sort out banana-trading policies that impact fairer trade. First, I briefly explain how the differing smallholder and plantation banana-production systems I described in chapter 2 led to divergent policies in Europe and the United States. This divergence helped distinguish African, Caribbean, and Pacific (ACP) banana exports from Latin American exports (sometimes called dollar bananas). Since the debate over trade systems represented an important political opportunity for banana stakeholders, I also discuss the process and resolution of banana trade issues that emerged in 2001 and their impact on banana workers and Fair Trade certification. Just how the new European tariff implemented in 2006 has affected Fair Trade choices I will consider in chapter 14.

Divergent Systems

While U.S. TNCs expanded their plantations in Latin America to ship bananas around the globe, certain European governments created alternative-marketing opportunities for smallholders in their own colonies. Britain, France, Spain, and Portugal each gave preference to imports from their possessions by offering inducements to encourage banana cultivation in often daunting island terrain. Bananas thereby became a major source of income and employment, helping to assure political stability. Via the Lomé Convention, signed in 1975, nine European nations assured forty-five ACP countries that the Europeans would maintain their banana industries, a commitment they periodically renewed until the end of the century.[1]

In 1993, the European Union struggled to combine the individual national import regimes into a single system.[2] Agricultural ministers from the United Kingdom, France, and Spain persuasively argued that, as the Lomé Convention specified, colonies and ACP countries should not be adversely affected by any new banana-import regime. Retaining their quotas would help preserve the smallholder culture and way of life without wreaking havoc on the land (see chap. 9). It also would retard drug production and immigration. By a narrow margin, countries that signed the Lomé treaty convinced the Union to maintain commitments by creating three banana-import categories:

- imports from current European territories, such as Spain's Canary Islands, Portugal's Madera, France's Guadaloupe, and Greece's Crete, which paid no tariff.
- imports from former territories in Africa, the Caribbean, and the Pacific, that is, the ACP bananas, which could ship at traditional levels

(up to 857,000 tons) tariff-free. Most ACP fruit was grown by smallholders, although its marketing was controlled by large companies.
- imports from Latin America. Usually produced on TNC-related plantations, these dollar bananas were assigned a 2.2 million-ton quota, which still accounted for the bulk of European imports. To balance the TNCs' lower plantation costs against the higher smallholder expenses, Latin shippers had to pay a 75 euro/ton tariff (adding about 7.5¢/lb.).[3]

The European Union also created an import-licensing system designed to encourage ACP banana distribution, with quota allocations from specific countries. Most licenses went to European companies that had produced or distributed from ACP sources.[4]

Considerable scholarly and political analysis has examined the impact of the new European trade regime, the backlash expressed through the about-to-be-created World Trade Organization (WTO), dominated by the United States, and the battle's severe effects on farmers and workers (see Sutton 1997; Grossman 1998; Raynolds and Murray 1998; Perillo and Trejos 2000; Fischer 2002; Myers 2004). The regime's justification gained positive support from the European public. However, it did not win over U.S. TNCs, especially Chiquita, which claimed that the new system brought a dramatic 65 percent drop from its prior share of the European market! Chiquita, Dole, and Del Monte Fresh Produce obtained most of their bananas from Latin America and felt squeezed by profit margins of 10 percent (Fairclough and McDermott 1999). Chiquita's profits had fallen from $172 million in 1992 to a loss of $52 million in 1993, and its share price had dropped precipitously.[5] Led by Chiquita,[6] the TNCs challenged the EU's populist claim that the new rules primarily benefited smaller producers in current and former colonies. Instead, they argued that the program unfairly raised tariffs while transferring shipping licenses to European-based firms (e.g., Fyffes, Geest) that quickly dominated the European import system.[7] Even though only 8 percent of Europe's imports came from ACP countries, the licenses guaranteed 40 percent of its market to European companies, substantially slashing TNC sales opportunities.[8] Effectively, the new system encouraged Latin exporters to purchase marketing licenses from traditional ACP/EU shippers. Since Europeans paid more for bananas, these licenses became increasingly valuable. TNCs began to buy and sell them on a grey market that was ripe for manipulation.

TABLE 7.2. TNC Percentage Share of Sales, European Banana Market, 1992–2007

TNC	1992	1996	1998	1999	2003	2004	2007
Chiquita	30	22	20	22	25	21.6–25	30+
Dole	12	15	16	21	17	13.1–17	12
Fresh Del Monte	5	13	16	18	15	9.4–15	14
Fyffes	4–5	19	18	2.5	7	20–20.4	7+
Noboa	7.5	7–10	7–10	9	11	5.8	

Source: Estimates drawn from data in *FruitTrop* 1999; *EuroFruit*; van de Kastelle and van der Stichele 2005, 15; company annual reports.

Dole and Del Monte played a double game as they maneuvered new license acquisitions, Dole increasing its European market share by four points between 1990 and 1994 to reach 15 percent (see table 7.2). Del Monte functioned more effectively with the regime after its change in ownership in 1996 (see chap. 2). However, it had "to pay high prices to place bananas in the European market, especially in the high demand season . . . [W]e spent $44 million in Euros to buy licenses in 2004 which we had to take from consumer pockets." Licenses also require "producers to sell the same amount they sold in the first half of the year, so the three largest companies are forced to toss a lot of bananas in the second half of the year."[9]

Although Chiquita secured more licenses, it chafed at past losses and envisioned more troubles ahead. Leading the others, it invoked Sec. 301 of the U.S. Trade Act of 1974, which required an official trade-impact study.

Declaring a Banana Trade War

In 1993, the major U.S.–based banana firms began a frontal assault on the European trading system that would have a lasting effect on global relations. They engaged the General Agreement on Tariffs and Trade (GATT) and the WTO to largely dismantle banana trade rules, forcing a redefinition of trade fairness. Responding to the TNC petition, U.S. Trade Representative (USTR) Mickey Kantor determined that European banana sales of the big three TNCs had been adversely affected by the

European system, in violation of GATT. Kantor expressed concern about job losses within the United States, where the TNC firms employed 7,500 in banana operations (USTR 1995). Other Latin American countries joined with Kantor to argue that Europe's new approach contravened its free-trade commitments.

Caribbean representatives disputed TNC and USTR claims, saying the U.S. TNCs had dramatically increased production in the early 1990s to gain greater European market share (see chap. 2).[10] They also argued that Chiquita experienced its lost profits and market share *before* the new system had been implemented and the company returned to viability in 1993. Since other firms had adapted to the trade regime, characterizations like Chiquita's "ignored the fact that ACP bananas were a great deal less profitable to import . . . cost twice as much at point of origin, cost more to ship. . . ." The representatives also stressed how much the regime's attack would threaten their vulnerable economies (Myers 2004, 78, 128).

Nevertheless, after hearings in 1993 and 1994, GATT review panels agreed with Chiquita and the United States that the Europeans were not in compliance with treaty obligations. This angered the Europeans, who prevented a final GATT consensus on the matter. However, Europe did agree to cut the import tax for non-ACP bananas to 75 euros per ton (Sutton 1997) and sought an official waiver for EU trade preferences.[11] It also gained a framework agreement with four of the hemisphere's major producers—Colombia, Costa Rica, Nicaragua, and Venezuela—that granted each country favorable import-tonnage quotas. Although the accord assured Chiquita of 25.6 percent of European banana imports, the fruit giant remained extremely dissatisfied.[12]

The GATT banana dispute came just at the time that the United States (with some European support) was engineering the creation of the WTO. The WTO replaced the GATT consensus process with the U.S. free-trade version of a "rule based trading system with teeth" (see Frundt 2005a, 221). Decisions of WTO panels could be reversed only by the unanimous agreement of *all* WTO participant nations, virtually impossible to achieve. The offended country could penalize countries found not in compliance for an amount equal to the lost trade. Chiquita then pressured the U.S. government to file suit under WTO regulations, arguing that penalties should be imposed on Europe to assure equity.

According to investigative journalists, Chiquita's $5.5 million U.S. political contributions enhanced this pressure.[13] The United States elicited the support of four Latin countries that had not signed the framework with the Europeans, that is, Ecuador, Guatemala, Honduras, and Mexico (and their independent producers), to invoke the WTO dispute-settlement mechanism. European officials countered that Europe's trade regime was WTO compliant. But in May 1997, the WTO panel ruled that the Union plan violated WTO provisions.[14]

As the banana case became the first to test WTO appeal mechanisms, the Europeans asked the WTO appellate body to reject the first panel's finding, claiming it overly reflected U.S. influence. Caribbean producers felt completely excluded from the proceedings. Nevertheless, in September 1997, both the WTO's appellate board and Dispute Settlement Body decided in favor of the United States, and gave the Europeans until January 1, 1999, to comply with their rulings.[15] According to U.S. Special Trade Negotiator Peter Scher, "The fundamental premise of the WTO . . . was that it would not be like the GATT; that countries could not simply block decisions . . . [T]hey would have to pay the price for not complying." Scher also stressed the "distribution and service interests" of the United States, whose companies "have for over 100 years worked closely with Latin American countries to export Latin American bananas to Europe" (Scher 1998).

Equitable trade demanded market protection for these companies! Scher even added a humanitarian rationale: the tragic situation in Honduras and Guatemala following Hurricane Mitch (Scher 1998).

However, the Europeans remained suspicious of U.S. intent, and of Chiquita as a million-dollar contributor to both Republican and Democratic political parties. The Europeans questioned U.S. government and TNC commitments to honor African, Caribbean, and Pacific protections, since, aside from allowing a differential ACP tariff, the United States had not offered any specific proposals or numbers. In fact, the United States had shot down every one of the European Community's own efforts to protect small ACP suppliers. TNC economists had chided Britain and France for encouraging inefficient ACP production that gained only eight tons of bananas per acre compared to the nearly thirty tons per acre that the companies achieved in Latin America. European representatives retorted by contrasting their objectives with the TNC's monocrop conversion of tropical rain forest and smallholder lands, and its extensive environmental destruction, exacerbated by spraying deadly pesticides (see USTR 1998a).

The U.S. Trade Representative characterized the European position as undermining "the viability of the WTO as a forum for resolving disputes . . . threatening the effectiveness of the multilateral trading system as a whole (USTR 1998b)."[16] The United States moved to unilaterally impose $520 million in penalties.[17] Once again, the Europeans appealed to have the decision revoked. Finally, in April 2000, the WTO's Dispute Settlement Body reduced the U.S. claim by nearly two-thirds to $191.4 million but said the aggrieved could now impose these penalties on imports from the European Union.[18]

So the banana case became a battleground for testing two different banana-trading systems—one claiming to defend small banana producers against U.S. TNC interests, the other claiming to stress a trading scheme equitably based on past markets. European officials firmly held that the United States was acting as spokesperson for the big U.S.–based banana companies, notably Chiquita, which appeared to dictate U.S. trade policy;[19] whereas the United States warned that under the guise of aiding former colonies and small producers, the Union was actually protecting its own banana-trading companies and middlemen. For example, Fyffes, of which Chiquita had divested itself in 1986, saw its sales triple between 1992 and 1997, largely owing to its European licenses. It also formed a joint venture with the Windward Islands Banana Development and Exporting Company to take over Geest's franchise. The United States argued that if the European Union prevailed, it would undermine the entire WTO process and force a return to the old, toothless GATT. President Clinton boasted, "When we face a judgment, we comply; the European Union seeks to change the rules." Yet French President Chirac retorted in their joint press conference, "We represent the workers" (Clinton 1999). No doubt unintended, the French claim offered an opportunity for fair-banana activists to question both U.S. and European trade regimes.

"First Come, First Served"

By the century's turn, the WTO's "nondiscrimination" approach had succeeded in gradually undercutting European efforts to support equitable trade with former colonies. In a final effort "to resolve" the banana dispute in the midst of a worldwide glut, European Union officials proposed a "First Come, First Served" plan in lieu of failure to agree on a reference period for estimating national quota allocations. The United States had earlier advocated a similar First Come, First Served system that claimed to allow those who possessed market-ready bananas to sell them where they wished. However, in oversupply conditions, TNCs had cut back on purchases from independents (see chaps. 2 and 8). Although the TNCs had begun to reverse this practice by fall 2000, the European Union's "First Come, First Served" proposal caught most of them by surprise. According to the proposal's terms, any western shipper with a committed cargo, including those in transit, could call ahead in any two-week period and bid for a proportionate share of the European Union's overall 2.5 million-ton Latin allocation. By eliminating national Latin quotas and licenses, the plan would stimulate a battle among Latin exporters in which losing shippers would have to pay a 300 euro/ton charge to sell bananas in Europe or peddle their excess to markets farther east at substantially discounted rates.

Whatever its intentions, the European First Come, First Served plan made few friends. Although struggling ACP suppliers would keep their quotas, they feared TNCs could adjust shipping schedules to dump overage that would depress prices (Myers 2004, 116). Some TNCs and independents spotted benefits, but others viewed the proposal as disruptive of past marketing arrangements (Raynolds 2003). Chiquita, already verging on bankruptcy because of hurricane losses, feared severe trade reductions. In January 2001, it countersued the EU for a claimed $525 million in losses. Most Latin exporters thought the EU plan would favor shippers that had faster boats and were closer to Europe, advantaging Dole, Fyffes, and independent producers with spacious transport that could procure the cheapest bananas.[20] Dole was in the best position to quickly ship to Continental markets. It had an ACP foothold, controlling 60 percent of the exports from Cameroon and Ivory Coast. Having terminated independent arrangements in Costa Rica and Colombia, it also sourced 35 percent of its bananas from Ecuador.[21] Even though its small and medium growers opposed the plan, afraid they would lose out to the larger national producers, the Ecuadorian government believed its shipments would double.

Latin banana unions estimated the First Come, First Served plan would jeopardize hard-won contract provisions. The thirty-five thousand workers represented by COLSIBA handled 40 percent of non-Ecuadorian bananas. If Chiquita folded, they would lose at least two-thirds of their members and weaken their negotiating power elsewhere. In a whiff of an early alliance, three advocacy groups—USLEAP, Global Exchange, and EUROBAN—gathered 172 signatures from major organizations to petition Dole to back away from the proposal. Fair Trade importers who themselves had scrambled to form consortia to bid on import licenses also understood the long-term drawbacks for banana producers and joined the request. Dole issued no reply. Claiming to act independently of Chiquita, Dole, or Latin countries, U.S. officials vehemently insisted that by preserving even minimal quotas, the new First Come, First Served plan violated World Trade Organization rules.

Temporary Opportunities and Resolutions: 2001–2006

Fresh trade alliances emerged following huge protests against the WTO meeting in Seattle, Washington, in December 1999. Shortly thereafter, European and ACP nations signed a successor accord to the Lomé Convention at Cotonou, Fiji, that made a somewhat weaker commitment to ensure "the continued viability" of national banana export industries (Cotonou Accord, February 2000, Article I). In lieu of trade preferences, they promised to establish reciprocal economic-partnership agreements (EPAs) which, assuming WTO permission, they would specify by 2008. New faces in Washington also encouraged a new approach. In April 2001 USTR negotiated an agreement with the European Union (and a second agreement joined by Ecuador) that accepted the 1994–96 reference period preferred by Chiquita for determining customary markets. European companies such as Fyffes lost market share. Preferential tariff access would remain for ACP countries until 2008, when the waiver for the Cotonou Accord expired and volume restrictions would be removed.[22] However, the ACP quota was reduced to 750,000 tons, and import guarantees for specific Latin countries were ended. Unionized producers in Central America won temporary relief, since the overall quota for Latin American producers would hold until 2006. The quota system would then sunset and a new tariff would be imposed. However, Windward producers were not pleased, and the accord made no mention of standards advocated by Fair Trade/FLO importers, who quickly wrote the European Commission seeking a more encouraging arrangement.[23] Dole saw the accord as a giveaway to Chiquita. Nevertheless, in exchange for the agreement, the United States suspended its sanctions on European imports that the WTO had finally permitted. In the end, while the United States and Europe preserved their separate spheres of influence, the two systems had drawn closer to a free-trade regime that took little account of smallholder livelihoods or worker rights.

During the 2001–2006 interim period, the TNCs retained an advantage over the independents, from whom they now sought shorter contracts. No longer having to deal with national quotas, the TNCs also manipulated one country (and union) against another. This inspired national competition based on product quality (see chap. 11 and COR-BANA 2004) but also promoted downright graft. In 2000 alone, more than 100 million pounds of Latin bananas had already entered Europe on forged licenses, further depressing prices (Myers 2004, 130). More subterfuges were anticipated. The TNCs also had to contend with the growing power of supermarkets, who made their own contract and monitoring arrangements and engaged in discount battles.[24] Del Monte underbid others to win Wal-Mart/Asda contracts in the United Kingdom in 2000–2002. Yet by 2004, Chiquita still dominated the European market and stood equal to Dole in world sales, as tables 7.2 and 2.1 demonstrate.

However, the new accord made things worse for smallholders in the Windward Islands, who had already suffered under the 1993 European regime. Their shipments to the United Kingdom declined from 238,000 tons to 99,000 tons in 2002 and the number of farmers dropped from 27,000 to 7,000 (NREA 2003)! As Anton Bowman, of the Windward Islands Farmer Associations, put it, "Before the WTO, we could get a subsidy from other countries; now we can't get this. After the trade change in 2002, the U.K. market was no longer guaranteed. We must fight to sell our bananas. It is going to make this region even poorer. They should realize that when a person is hungry, he can be very angry . . ." (interview with Bowman 2005). Yet to no avail, ACP nations petitioned Europe for higher quota protection (Myers 2004, 138).

Between 2001 and 2006, the agreement did stimulate new thinking within the Latin region. The guaranteed quota offered breathing space to prevent further union erosions, but enlightened leaders realized they had little time to forge a new strategy on the ground. Although Europe received 69 percent of its production from Latin America, as opposed to 10 percent from Africa, 5 percent from the Caribbean, and 16 percent from its own domestic producers, the Latin and Caribbean portions would soon decline.[25] Opportunities were ripe for farmers, workers, and activists dismayed over trade inequities to seek common purpose. In 2005, unionists and island producers organized a second major conference to prevent the banana race to the bottom, and other rallies followed.

Fair Trade's Own Trade Standards

Fair Trade represents an effort to redress trade inequity. To rebalance the relations of power, FLO certifiers insist that exporters, importers, distributors, and retailers abide by special rules that assure a fair marketing process. These are additional to Fair Trade's criteria for social, economic, and environmental protections outlined in earlier chapters. FLO thereby expects that producers will minimize wild price swings so they can implement plans for lasting environmental and labor improvements.

To achieve a rational approach to marketing, FLO promotes direct purchases so that an importer and certified producer can establish a lengthy relationship. FLO prefers that the two parties sign a contract that sets quarterly production and purchase targets spread over at least a year. The contract also specifies tolerances for product quality, packing methods, delivery conditions, liability, and price and payment terms. Divergences are not permitted unless listed in the contract, with a copy submitted to FLO. Importers are obligated to buy at least half of the amount a producer projects to produce over a three-month period. Ordinarily, importers confirm purchases two weeks prior to their receipt of an order. Importers are then expected to renew orders for the next quarter unless they inform producers with a two-week lead time. An exception would be if more than 15 percent of the bananas were not up to standard. In such a case, they could still be sold via the contract under a special clause. The importer could take them on consignment, promising the best possible price after deducting a 6 percent maximum commission.

While the two-way arrangement is preferred, FLO does allow more complex trading agreements to make better use of shipping schedules and container space (FLO 2006). As long as Fair Trade principles are honored, FLO may approve exporters that pay producers a "farm gate" price that does not include the additional costs of the "freight on board" price. Nearly always, however, FLO determines the floor box price for each country. While this price is affected by the production costs within that country, in no case does it dip below the general floor price that FLO sets every two years. In 2002, for example, it was $5.25/18.14 kg box. If the conventional market price rises above this floor, FT producers receive the higher price.[26] Nevertheless, the reality of FLO pricing is not always so smooth, as Shreck (2002; 2005) and others have documented. Complications can arise at the local level as exporters purchase proportionate quantities of fruit, some of which cannot be readily sold and some of which does not measure up to Fair Trade standards. In Shreck's random sample of Finca 6 producers during the economic downturn in 2000 in the Dominican Republic, 78 percent said the FLO price did not cover their costs.

Besides the price question is the anticipated market. Most bananas sold in the European Union—Fair Trade or not—are handled by distributors who had been granted official licenses to sell in certain areas. As noted above, in the 1990s these licenses became increasingly controversial, but they did offer one way of rationalizing wholesale distribution. In entering the game, Fair Trade/FLO had to jockey for sometimes costly license arrangements that affected both the amount sold as well as the price. Both smallholders and workers had a direct interest in distribution arrangements. When volume fell below target production or price, their livelihoods were jeopardized. African Fair Trade unionist Gariba Musah warned that as long as the price was reasonable, "people should buy our bananas because they are fairly produced. We workers here in Ghana can also benefit from the export and get something. Volta River Estates provides employment opportunities for this area. It is very difficult to find work here" (Liddell 2000, 15). But as European licenses became costly or unavailable, Fair Trade producers suffered along with unionized producers.

COLSIBA affiliates as well as island producers seek equitable trade rules to balance quota losses. George DeFreitas, Nioka Abbott, and other committed small farmers remain confident in FLO protections. If all goes well, they are paid within two days of port arrival. If not, they get at least 25 percent of payment within a month. In case of shortfalls in sales, they may have to take a 10 percent loss, but then the importer assumes the rest of the loss. Since FLO offers to arbitrate disputes, this system usually works. However, if world prices suddenly rise, TNC shippers may front funds to local purchasers and undercut FLO's ability to rapidly respond, potentially restraining its budding effort at fairer banana trade.

Summary

I have demonstrated in this chapter how farmers, unionized workers, and their supporters have networked to ply the competing structural opportunities of trade. Via the Lomé convention, smallholder associations gained notable benefits that were at least partly enshrined in the EU trade regime. When the United States challenged that regime in the 1990s, the Europeans battled U.S. and Latin corporate interests, but peasant and union alliances drew closer to conceptualizing a mutual pursuit of class-based trade policy. Although ultimately unsuccessful, they developed closer linkages and identities as a Fair Trade alternative vision emerged. Before discussing the potential of labor and smallholders to form a transnational activist network, however, I must consider their separate historical confrontations with the dominant systems of banana production.

CHAPTER 8

Resilience of Banana Unions

I HAVE ESTABLISHED that corporate banana systems, environmental conditions, and trade policies offer structural opportunities for common understandings about fair bananas. The key question we now face is whether potential alliance participants can forge linkages and identities. In this chapter I briefly summarize the history of banana-worker struggles, recounting how and why field and packing hands mobilized networks and asserted their right to unionize and bargain collectively. I show how, in parallel with small farmers, they staked their own claim as qualified monitors of nature. As these workers expressed elements of collective identity, unions became early supporters of Fair Trade, and FLO participants gradually grew to appreciate the importance of employee rights.[1] In turn, consumers chose to purchase Fair Trade bananas as a means to encourage respectable treatment for workers, as well as for farmers and nature.

Banana workers arise before daylight, then walk or catch the bus to their assigned field or packing tasks. Some cut weeds from the irrigation channels, or prepare cultivated areas for new plantings. Specialized field hands cover the growing fruit with plastic bags and perform spot spraying for nematodes. As the day's heat intensifies, teams of three or four can often be seen standing in potentially toxic or snake-infested water cutting ripe bananas. Strapping the 100-pound stems to their backs, they deliver the stems to a cable line as much as a half-mile away. At the packing plant, a largely female labor force unhooks and cuts the clusters into bunches. As an automated cable moves the fruit past workstations, repetitive cutting and sorting can cause traumatic stress. Women like Auria Vargas wash the bananas, often in chlorine baths, sort and spray them with fungicides, then pack the bunches in forty-pound boxes for shipment. Chemical exposure often elicits rashes, and at times induces vomiting and permanent cancers. Sanitary facilities are routinely appalling, with workers having no access to functioning toilets or potable water.[2] Field and plant workers commonly put in ten- to fourteen-hour days. When paid by piece rate the women workers feel more vulnerable to harassment by male supervisors. The women also express concern about the lack of care for their children. Since there are no windows to close against drifting aerial sprays, all face potential toxic contamination when they return home. Their water may not be drinkable, and washed clothing may contain pesticide residues. In sum, owing to such conditions, banana workers are susceptible to high rates of sickness and death (Longley 2005).

When laborers first complained about such conditions, United Fruit and other company officials dismissed their demands. The company imported "docile" laborers, playing migrants against residents and one ethnic group against another to gain a wage advantage (e.g., Purcell 1993). Yet as market conditions and the political environment improved, workers took opportunities to organize strong unions throughout Central America and Colombia that banana TNCs learned to tolerate. Yet with the third wave of banana expansion in the 1980s, labor's opportunities dwindled as the TNCs reversed their more enlightened practices. Starting in Costa Rica, they accelerated their attacks on independent worker organizations, especially as global banana supplies increased following a devastating hurricane in 1998. In this chapter I review how worker networks expanded in various countries. I then discuss the way rising union identity became linked with environmental issues, how independent employee organizations confronted fresh repression, and why those pursuing fairer trade added criteria that emphasized worker rights.

Unions Arise 1910-1960

In its early years United Fruit officials appeared oblivious to the strenuous and risky work of their employees. The company paid its workers little and strongly repressed labor organizing. Yet field and packing hands struggled for dignity in each country where the company expanded. By 1918 United Fruit had gained a bitter reputation for opposing unions and strikes and for creating divisions among employees in Colombia. Within ten years however, the banana workers of Magdalena joined together with farm laborers, rail workers and dockworkers, *and* land squatters to fight for land and water rights. The coalition proved such a threat that the company solicited the government to send in troops. They arrested four hundred strikers and fired into a huge crowd, killing hundreds. Yet disaster spawned opportunities. A more sympathetic government recognized a newly formed agricultural union six years later. United Fruit halted further investment and prevented formation of a communist-led union. Decades later, in 1964 SINTRABANANO (Sindicato de Trabajadores Bananeros [Union of Banana Workers, Colombia]) struggled to organize United's Frutera de Sevilla affiliate and likewise endured vehement retaliation. As sigatoka became widespread, the TNC reduced its workforce and claimed to be leaving the country (Pedraja Toman 1987, 202–4). In 1972, the Christian-based union of agricultural workers, SINTAGRO (Sindicato de Trabajadores Agricolos [Union of Agricultural Workers]), started organizing banana laborers. During the decade SINTAGRO and SINTRABANANO competed to recruit the largely Afro-Colombian workforce in Urabá, the primary region for producing export bananas. When Standard Fruit arrived in 1977 the growers invited a heavy military presence, restricting genuine contracts to 10–20 percent of the workers (Chomsky 2004b). During its third wave of expansion in the 1980s, United returned to Colombia and accepted a unionized workforce, as I elaborate below.

As their cross-border collaboration increased during the 1920s and '30s in other nations, workers further exploited opportunity structures for organizing. Unions achieved success in Costa Rica in 1933, where the local Communist Party solicited Atlantic-zone banana workers to join a massive strike against United. The resulting Confederation of Costa Rican Workers pursued the company when it shifted growing operations to the Pacific coast. During the 1940s and 1950s, the confederation created a network of allies, including Christian unions, to solidify its strength among fruit workers. Over the next thirty years, militant Tico (Costa Rican) banana unions gained in wages and benefits. However their unification efforts in 1980–81 and a sixty-seven-day strike at Del Monte led to a union dispute over communist influence (Booth 1998, 273). TNCs grabbed the opening to replace unions with management-labor "Solidarista Associations."

In Panama, banana workers had attempted to organize on both coasts since the 1930s. Each time, United Fruit "engaged in massive firings and eliminated all the organizers. One tactic . . . was to load all 'rabble-rousers' with their families and belongings into trains, take them deep into the grasslands of Chiriqui, and abandon them" (Phillipps 1987, 591). Not until a major strike in the 1960s did UF officials allow space for union organizing. Union leaders also gained status under the Torrijos government (1968–81), inspiring United Fruit to lease more lands to independent growers.

When workers at United Fruit–Guatemala invoked the new labor code in 1948 and 1949, a U.S.–supported company union armed itself to attack. Government efforts to repossess unused UF property further precipitated calls for foreign intervention (see chap. 2). The U.S.–engineered coup in 1954 unleashed mass slaughter and union repression that even U.S. officials eschewed. Most worker organizations ceased to exist, and United Fruit workers took cover within a U.S.–backed confederation for agricultural workers. When United Fruit sold significant properties to Del Monte in 1970, its Guatemalan union re-emerged as SITRABI (Sindicato de Trabajadores Bananeros de Izabal [Banana Workers Union of Izabal]), currently the nation's oldest existing union. SITRABI never became a politically militant force (González 1988), but it gained solid contracts with Del Monte that provided housing, sports facilities, schools, transport, health services, and a livable wage (Frundt 1995).

When bananas became Honduras's principal source of foreign exchange in 1902, the government surrendered land concessions that gave United Fruit, along with Cuyamel Fruit (later sold to United) and Standard Fruit, virtual control over its north coast properties. Historians argue that sufficient competition remained so that independent producers still could gain a suitable livelihood (Euraque 1996; Soluri 2005). Yet when the larger companies faced standoffs with their employees, the Honduran military would come to their rescue. In 1916, troops rushed to quell an early strike protesting Cuyamel company-store policies. More radical unions rose in the 1920s, angered over company attributions about "outside agitation" (Euraque 1996, 37). The government again deployed forces in 1932 when workers halted production to prevent United and Standard Fruit from reducing wages. However, in May 1954, fifty thousand workers from United, Standard, and related companies inaugurated the famous sixty-nine-day "Great Banana Strike" that shut down half the country, caused substantial company losses, and precipitated the formation of the Federation of National Workers of Honduras (Argueta 1995). Given what was happening in Guatemala, U.S. officials attempted to co-opt the strike committee, and advised United Fruit to negotiate with the noncommunist SITRATERCO (Sindicato de Trabajadores de la Tela

Railroad Company [Union of Workers of the Tela Railroad Company, Honduras]). However, when Hurricane Gilda's arrival led the company to fire ten thousand employees, SITRATERCO tripled in size, stimulating other unions and ultimately, a national labor code.

United Fruit workers of SITRATERCO constituted Honduras's leading union, followed by workers at Standard Fruit (Meza 1997). Although the United Fruit workers had affiliated with the right-wing, U.S.–supported American Institute for Free Labor Development (AIFLD), AIFLD's local Honduran office proved unusually flexible. Leftists gained influence and AIFLD unions grew militant.[3] When worker takeovers threatened Standard Fruit after the Honduran coup in 1975 (chap. 2) the army incarcerated two hundred strikers and installed a company union. Standard/Castle & Cooke "made regular payments to Honduran military officers." The army "handpicked" union leaders at a legally recognized worker-run plantation and forced "them to sell all of their production to Castle & Cooke." Workers still resisted, typified by one fired laborer: "We have suffered heavier blows from the banana companies and recovered. We will do the same again" (Burbach forthcoming). In 1980, Standard Fruit workers held a lengthy strike over higher wages and overtime pay (Pearson 1987, 472ff). Women likewise took the opportunity to gain their own committee and positions in the union sector (Frank 2005).

Ecuadorian banana workers organized when the government promoted banana-export development after World War II. When United faced the devastating Panama blight in the late 1950s, it cut wages and fired hundreds. Workers organized a cooperative that took control of a United plantation in 1962. When the government intervened with an agrarian reform plan that supported co-ops, the fruit giant left the country (Striffler and Moberg 2003; Larrea Maldonado 1987).

In 1925, Nicaraguan soldiers killed banana employees at Cuyamel Fruit when they petitioned for better conditions and a raise (Amador 1992, 60). The following year the noted revolutionary Cesar Augusto Sandino aided the UF struggle after contesting for control of the government, but subsequent opportunities proved hard to come by. In the mid-1960s workers organized as the Confederation of Union Unification (CUS), forming a relatively conservative network that functioned quietly under President Anastasio Somoza (1967–79) with AIFLD support. It eschewed involvement in the Sandinista uprisings.

Thus, by the 1960s, banana workers in Colombia and Panama had gained sufficient protective legislation to merit what labor analysts call the "Fordist model" of labor relations in which workers receive adequate pay for work, albeit arduous and regimented (Harari 2005b).[4] Honduras and Costa Rica had also made advances, with Ecuador, Guatemala, and Nicaragua trailing behind.

The Attack Against Unions 1980-2000 and Beyond

Banana union successes during the 1960s and 1970s inspired numerous company stratagems to undermine union strength, from creating in-house unions to making "protective" arrangements with local "independent" producers. Most of these efforts failed. However, when banana companies again expanded in the early 1980s, militant unions proved the major obstacle, and the TNCs decided to take a stand: "I got the call from our headquarters in San Francisco to 'cut loose' from the unions and take our losses," reported one Dole manager. "This is what I did, and just in time too" (interview with Arana 2001).

In Costa Rica, the companies cut loose by creating a sophisticated legal alternative to unions called a Solidarista Association. While appearing employee-oriented, companies could manipulate Solidaristas to avoid what unions offered and Fair Trade would later demand. "The banana TNCs began an intense antiunion campaign that gained support from intellectuals who believed that banana workers had grown too combative and communist party–oriented; and from conservative church leaders who advocated Solidarista Associations," complained Gilberth Bermúdez of COLSIBA (interview with Bermúdez 1999). Chiquita produced a Solidarista video that the John XXIII Center promoted. "Ninety percent of our twenty-five thousand workers were unionized in 1982, but with such tacit backing, corporations won changes in the labor code that reduced union protections" (Ibid.).

The newly changed rules encouraged companies to create permanent committees that could sign direct agreements (*arreglos directos*) with each finca. Such arreglos directos had no legal standing to guarantee banana workers basic protections over wages, benefits, and job security (Wedin 1986; Flores Madrigal 1993). Yet because they could offer loans

FIGURE 8.1. Negotiated worker housing, Sixaola, Costa Rica. (Photo by author)

to workers, company managers presented them as an enticing option. If TNCs could cajole five employees on each finca to sign an arreglo, they could then prevent genuine negotiations. Costa Rican banana workers boldly resisted the changeover with several lengthy strikes. However, news media portrayed the strikes as socially detrimental, and companies repressed union leaders with massive firings. Thereafter, employers felt they could disregard labor laws with impunity. They refused to increase wages, consulted blacklists, subcontracted, and rotated workers on a three-month basis to avoid benefits (Hermosilla 1998).

Elsewhere, TNCs were less successful at introducing Solidaristas. In Guatemala, the 4,300-strong SITRABI union of Izabal earned twice that of other area workers, had a well-functioning grievance procedure, and enjoyed subsidized benefits. But in 1987, when management assigned a vacated union position to a solidarista, SITRABI conducted a twenty-nine-day strike without pay. "Our members viewed the strike crucial to prevent further erosions," explained Committeeman Marel Martínez (interview 1989). Rank and file subsequently praised the union for bringing "improved conditions in health and education and truly representing the workers' interests" (Frundt 1995, 295; see fig. 8.1.)

Congruent with SITRABI on Guatemala's Atlantic coast, UNSITRAGUA (Unión Sindical de Trabajadores de Guatemala [Union of Independent Workers, Guatemala]) organized eight Chiquita fincas[5] and confronted the TNC's practice of leasing its lands to antiunion independent producers via dummy corporations to avoid high labor costs. When UNSITRAGUA petitioned for the workers' legal rights, Chiquita officials collaborated with the independents, who summoned police to arrest UNSITRAGUA leaders as they attended court on the workers' behalf (see Perillo 1998).

Honduran, Nicaraguan, and Panamanian unions likewise resisted TNC manipulations. Honduran workers fought United Fruit's efforts to diversify banana lands into African palm production (see Coor 1999). In 1981, after the Sandinista Revolution, Nicaragua gained an agreement with Dole that gave the government administrative control over company fincas. Banana production returned and some European women's groups promoted revolution bananas. Nicaraguan workers joined the militant Association of Rural Workers (Asociación de Trabajadores del Campo [ATC]) and the Federation of Banana Workers (Federación de Trabajadores Bananeros de Chinandega [FETRABACH]). Dole offered technical assistance via five-year contracts until it departed in 1999. In 2008, the ATC sought to negotiate sales to Chiquita.

Panamanian strikes in the late 1970s and early 1980s elicited United Fruit threats that it would close. By 1998, employees at Armuelles, on the Pacific, had reached their limit after years of flat wages. They struck again for nearly two months, stating their breaks remained uncompensated and a new packing process had brought cuts in pay. They charged that export production was being redirected to the Atlantic coast, costing 240 jobs.[6] United claimed to lose $90 million due to the strikes and irrigation problems (Taylor and Scharlin 2004, 208). It sold its twelve Armuelles fincas to union employees in 2003. They became a Multiple Services Co-op (Cooperativo de Servicios Multiples de Puerto Armuelles, COOPSEMUPAR) with a ten-year, exclusive commitment for production and sales but also retained their identity as the Union of Workers of Chilco (Sindicato de Trabajadores de la Chilco, SITRACHILCO). Workers in the Industria Bananera (SITRAIBANA) union at Bocas del Toro, on the Atlantic, refused any buyout, as did the workers in nearby Sixaola, Costa

Rica. In 2002 SITRAIBANA walked out for eleven days to protest lost hours following postharvest mechanization.

In Colombia in the early 1980s, as guerilla forces expanded in the Urabá banana region, they supported labor struggles, and gained an accord with the government.[7] This stimulated both Standard and United Fruit to divest themselves of direct property ownership but to retain informal control. Nevertheless, they could not prevent an industrywide strike in 1984 that won 127 local contracts covering 60 percent of banana workers. However, in 1987 a new national administration abrogated its guerilla accord, and paramilitaries infiltrated the area. Although unions achieved an eight-hour-day contract for 85 percent of the workers, scores of rank and file were assassinated. In 1988 the various unions joined forces as SITRAINAGRO (Sindicato Nacional de Trabajadores de la Industria Agropecuaria [Union of Workers in Industrial Agriculture, Colombia]); however, landowners and politicians played the two major guerilla groups, ELN (Ejército de Liberación Nacional [National Liberation Army, Colombia]) and the FARC (Fuerzas Armadas Revolucionarias de Colombia [Columbian Revolutionary Armed Forces]), against one another, unleashing an unprecedented rampage against labor activists. While SITRAINAGRO's monthlong strike in 1989 accomplished a hefty wage and housing increase, the banana union gained no human-rights assurances and lost nine leaders to assassination.

COLSIBA and Its Contestations

In 1992 struggling banana-union leaders in Costa Rica, Honduras, and Colombia established the Coordination of Latin American Banana Unions (COLSIBA) to resist corporate and governmental attacks. By 1998, the COLSIBA network had won beneficial contracts in Guatemala and Panama as well as in Honduras and Colombia, where most banana workers were organized. Following a "transnational activist" campaign in early 1998, Costa Rica's four COLSIBA unions also developed a unified program. Aided by a U.K. campaign that dumped banana skins outside Del Monte warehouses (Lamb 2008, 12–13), one Costa Rican union signed an accord with Del Monte/Bandeco in which the company endorsed "respect for free unionization," allowed organizers access, and "the right of the worker to his/her free election of choice."[8] Two unions in the Sixaola region negotiated with Chiquita, and another organized near Nicaragua. "We have a much stronger base with the companies," boasted Bermúdez (interview 1999). Union bulletins questioned whether company certifications improved social conditions: "Field hands still worked long, difficult hours, often lasting from 4:00 a.m. to 6:00 p.m." (interviews with Workers of Sarapaquí 1999).

Five thousand women among COLSIBA's thirty thousand members earned the union's attention. Averaging "3–4 children each, we face issues regarding child care, domestic violence and health problems. In Honduras 400 women have died in work-related activities and the rate is even higher in Guatemala," emphasized Iris Munguía, COLSIBA's secretary for women. As women increased their union participation and gained collective pacts, Munguía believed, "[W]e avoid many of the violations of basic rights. These include maternal time off for feeding . . . equitable salaries and hours" (2005). Munguía and other union women demanded that health concerns, sexual harassment, and other family issues be addressed (Frank 2005).

Despite some successes, by 2000 the Costa Rican unions had not signed any contracts outside Sixaola. Severe housing problems remained. One Del Monte finca furnished only fifty houses for eighty families, thirty bunks for sixty-five single men, and no housing for contracted workers. The company persisted in spraying over living quarters. Collective negotiations had gone nowhere. If workers tried to mobilize a new union, they were often assigned less remunerative work and sometimes fired. In effect, Del Monte had undermined its agreement and "the Solidarista Arreglo had no legal force either. We can't even arrange a meeting to discuss it, while the company has the help of the John XXIII [Solidarista] Center, which favors management" (interviews with Workers of Sarapiquí 1999).[9]

COLSIBA's role in Colombia deserves thoughtful consideration in light of United Fruit's return to Urabá in 1989 for reasons similar to those explaining its expansion in Costa Rica. At that point the competing FARC and ELN guerillas demanded protection money from Chiquita, creating deep factions within SITRAINAGRO's eighteen thousand workers. Internal guerilla conflicts resulted in forty employee deaths in 1993, precipitating a massive worker demonstration in the local sports stadium.

By 1995 hundreds more had been killed, but a good percentage were likely murdered by the paramilitary AUC (see Chomsky 2004b). In chapter 2 I described the government's efforts to establish Convivir "cooperatives" as an AUC front to secure the region. Forty thousand Colombians marched to protest the resulting kidnappings and killings in 1996.[10] However, as the AUC wielded the carrot of a guerilla extermination and the stick of fresh death threats, it gained the allegiance of Chiquita/Banadex officials, along with that of other companies and some unionists. Chiquita switched its protection payments from guerillas to the AUC the following year.

Such protection did not always work, as "Alberto" and co-workers vividly remembered when they arrived for work at "Banafinca" in 1999 and saw two AUC

> henchmen standing menacingly near the packing plant. "No one knew who they had come for that day," Alberto says. The thugs waited until everyone took up their workstations then went into the field where one of Alberto's coworkers was climbing a ladder to bag a banana stem. "They cut off his head with a machete, dumped the weapon, then calmly walked to their motorcycle and drove off, without saying a word." This was only one of ten murders that Alberto has witnessed on Chiquita fincas. It is his view that the company's contributions to the paramilitary groups helped strengthen them and allowed them to expand throughout the country. "The money Chiquita paid helped finance the paramilitaries. Their coffers grew, and they were able to buy more weapons." (Brodzinski 2007)

As the AUC neutralized militant labor activists, fear and pragmatism motivated SINTRAINAGRO's new union leadership to work more closely with government and grower bodies, including the military and, in some cases, paramilitary forces. In doing so, they elicited improved contracts and funds for worker housing and education. But SITRAINAGRO has been criticized for its silence, if not its direct paramilitary collaboration in order to achieve peace, increased production, and benefits (Chomsky 2004b). The reality is complex, as Chomsky acknowledges. Given the horrific struggle and leadership losses of the union, it sought a new approach in Colombia. Some courageous leaders still protested paramilitary payoffs, despite risks. Labor officials like José Benítez demanded that Chiquita and the other firms be held accountable for their complicity in assisting killers. "It's like they are trying to erase all those deaths with money that the victims here will never see. If there is justice, the Chiquita executives will see the inside of a Colombian prison," stressed Benítez (Brodzinski 2007).

Following the change in 1997, COLSIBA worked strenuously with limited resources to restore the union's independent voice and quell internal divisions.[11] SITRAINAGRO minimized its COLSIBA participation, but Danish, Finnish, and Spanish unions assisted with training and educational projects, hoping to instill a new approach. Finally, SITRAINAGRO joined the IUF. It then adopted a more integrative strategy downplaying political differences, and focusing on negotiations. A one-day strike in 2002 brought banana workers more funding for housing. Colombia's important CUT (Central Unitaria de Trabajadores de Colombia [Unitary Workers Confederation, Colombia]) endorsed SITRAINAGRO's strategic shift from political to social and economic priorities, and in 2006 CUT's general secretary praised the union's collective agreement and IUF affiliation as a model for the industrial sector.

Unions Adapt an Environmental Identity

Despite national differences, COLSIBA was able to build unity around environmental concerns. Although the union movement, like the Fair Trade movement, was not originally focused on ecological issues, the unions had learned from mistakes in the 1940s, when they did not prevent the spraying of the deadly methyl bromide (Marquardt 2002). In the 1990s, they took a more proactive position, contrary to a perception cited in Taylor and Scharlin (2002, 114) that workers were not convinced of the importance of conservation. COLSIBA articulated an environmental commitment at its first congress in 1993.[12] Affiliate unions regularly conducted investigations and seminars about pesticide contamination led by such leaders as Carlos Arguedas Mora (himself sterilized by DBCP), Iris Munguía, and Ramón Barrantes (e.g., Arguedas Mora 1999, 2000; Interview with Arguedas Mora 1999). In mid-1997, the Costa Rican unions publicized their "Social and Environmental Clauses for Banana Production," decrying sterilizations, sickness, and death from "DBCP in the 1980s." They proposed "international rules" for fumigation, storage, protective gear, and the "application of agrochemicals in all banana activities" (COSIBA-CR 1997). With help from a Danish labor federation, COLSIBA initiated a project on "Environmental Labor Conditions of Field Workers" throughout Central America.[13] Acknowledging the difficulties of training workers to understand chemical risks, COLSIBA offered technical sampling training for monitoring pesticides. Thus, in the space of a decade, with transnational activist support, unions in the COLSIBA network added the monitoring and prevention of environmental destruction as part of their identity. In May 2002 two thousand SITRATERCO workers struck fourteen fincas to protest the firing of twenty-nine workers who had halted their spraying of chlorpyrifos (Durban), fearing nausea, light-headedness, and sterility (see chap. 12). United/Chiquita agreed to modify the spraying arrangements.[14]

The Big Challenge

Weather in October 1998 dealt a shattering blow to banana production in Guatemala, Honduras, and Nicaragua, while largely sparing Costa Rica. Hurricane Mitch doused the region for five days with rain that measured in feet, not inches. Reporters described it as "the hemisphere's most devastating disaster of the century, one that turned rivers into raging torrents and unleashed landslides, killing and burying its victims all at once. It cost an estimated 9,000 lives and more than $9 billion in damages" (Thompson and Fathi 2005, 9). Workers crafted palm-canoes to escape the rising waters, but the extensive flooding, arguably made worse by monocultural land clearances, wiped out banana fields and washed away homes.[15] Companies considered terminating banana operations in the affected nations and quickly sought supplies of bananas from Ecuador (see chap. 2). Banana prices remained solid through most of 1999, as the market reached a balance. The TNCs reluctantly agreed to rebuild areas hit by Hurricane Mitch, in part because of pressure from unions. But then in late 1999, as low-cost bananas saturated world markets, demand peaked and organizing opportunities evaporated (Perillo 2000). Companies demanded severe labor concessions.

In Honduras, Chiquita insisted that SITRATERCO, the union that had inaugurated the national labor movement, modify its contracts and accept layoffs. In Nicaragua, Sandinista unions struggled as Dole, the only TNC investor, left the country. Panamanian workers barely survived their lengthy strike in 1998. In Colombia, where the internal war was taking its toll (see Chomsky 1996), battered banana unions faced demands for higher productivity as TNCs curtailed purchases. In 2002 in Ecuador, despite the enormous increase in production by national producers, the government allowed employers to quash organizing among formerly unionized workers and violently evict unionized strikers (see chap. 13).

Dole and Del Monte laid off as many as four thousand workers each in Costa Rica, and closed less-productive plantations. Del Monte actually fired, then rehired its employees, modifying the contract by extending work hours, refusing coverage for domestic electricity and medical costs, and cutting pay by 40 percent. Although Chiquita took measures that were less draconian, it notably lowered salaries. In the midst of negotiations, it fired its only officially recognized union leader in the country, and refused further discussion (COSIBA-CR 2000). Eighty-five percent of the company's workers remained in Solidarista committees (Chiquita Brands 2000). Chiquita and the other firms also rotated illegal Nicaraguan immigrants for ninety-day periods from one plantation to another to avoid salary and benefits requirements (Foro Emaús 1998a; Interview with Barrantes 1999). If supervisors could not terminate union workers outright, they assigned extra tasks for the same pay (interviews with Workers of Siquirres 1999).

In Guatemala, hired goons vehemently attacked the stalwart SITRABI, representing Del Monte workers in 1999 (fig. 8.2). Del Monte/Bandegua had just subleased three major plantations in the Bobo district and terminated more than nine hundred workers. As SITRABI planned a legal work-stoppage in response, it suddenly faced a squad of two hundred armed thugs encouraged by several of Del Monte's independent producers and led by city councilmen worried about potential trade losses. Del Monte henchmen then carried out their gruesome 1999 attack, dragging SITRABI leaders from their beds in the middle of the night and threatening them with death if they did not call off job actions, renounce the union, and accept lower salaries and layoffs.[16]

However, the union quickly notified the USLEAP, EUROBAN, and other transnational networks, which coordinated an intense international campaign that brought Del Monte to sign an agreement with the IUF (USLEAP 1999). It was a major victory for SITRABI, which gained a master contract for more than 600 workers.[17] The U.S. trade represen-

FIGURE 8.2. Del Monte SITRIBI union committee under threat, 2002. (Photo by author)

tative even announced a potential cut in Guatemala's trade benefits if it did not pursue the case. Finally, in March 2001, the union leaders faced their attackers in an open trial—a first for Guatemala. Remarkably, the legal system found some of the perpetrators guilty, although of lesser charges that did not involve incarceration. The SITRABI executive committee still went into exile for protection. Remaining leaders like Noe Ramirez and Selfi Sandoval had to pick up the pieces: "After the exile of our people, the problems persisted. During September and October 2001, we received phone calls in the middle of the night threatening to kill us. We filed a complaint with the Public Ministry but they sat on their hands."[18] SITRABI faced a new round of attacks following the killing of the head of another union in Puerto Quetzal in January 2007 and military presence at its own headquarters in July. In September, masked men murdered Noe Ramirez's brother, Marco Tulio Ramirez, SITRABI's secretary of culture, as he was leaving home for work on the plantation. In November, attackers shot at union officer Cesar Guerra; in December, at the house of Carlos Mancilla, appointed by SITRABI's

FIGURE 8.3. Unemployed Chiquita contract workers occupying closed Alabama/Arizona packing plant, Guatemala, 1998. (Photo by author)

confederation CUSG to investigate Marco's murder. In March 2008 they fired more bullets into the home of CUSG's general secretary. Added to other killings, these acts enraged the Guatemalan human rights community (see ACILS 2009).

While Del Monte's reaction to Hurricane Mitch went beyond the pale, the storm also served to justify Chiquita and Dole's displacement of Caribbean coast workers in order to plant African palm, which employed only a tenth of the workforce (Coor 1999; Trejos 1996). A Chiquita contractor abruptly closed his Alabama/Arizona plantations, precipitating an eighteen-month worker occupation backed by UNSITRAGUA (fig. 8.3). The TNC set up nonunion banana operations on Guatemala's Pacific coast, where pay and work conditions reflected those in Ecuador.

Unions strenuously resisted the abrupt loss of protections they had gradually won over fifty years of combative struggle (see the comprehensive AFL-CIO study edited by Perillo and Trejos [2000]). In addition to the Del Monte victory, Chiquita finally acquiesced to hire 110 of the Alabama/Arizona workers on its own lands, in part because of international pressure. Nevertheless, despite assistance from the IUF, EUROBAN, USLEAP, and other groups, labor's transnational activist effort could not alter the market glut. In watching Fair Trade proponents devote considerable resources to securing European market shares for small and independent producers, banana unions realized they also had to act for themselves. They demanded that the Fair Trade campaign pay attention to worker conditions (interview with Bermúdez 1999).

Fair Trade and Social-Development Standards

The unions saw that what separated the Fair Trade alternative from most other North-South (or South-North) programs was its insistence on a sustainable livelihood for producers and workers, even in the face of hurricanes and oversupply. It demanded that a just banana system fully honor labor rights as well as environmental principles. As FLO organizations refined their social-development standards they sought to aid the socioeconomic development of hired plantation workers as well as of independent producers. They turned to core policy agreements (Conventions) developed over ninety years by the International Labor Organization (ILO).[19] For example, FLO incorporated into its own standards ILO Convention 111, which prohibits discrimination and requires employee organizations to exert strong efforts to assure proportionate participation by those gender and ethnic groupings present.

Union leader Gariba Musah, whom we met in chapter 3, expressed enthusiasm about FLO certification because it guaranteed employment at a reasonable wage for the Ghana Workers Union (GAWU) at Volta River Estates, Ltd. (VREL) one of the country's two certified plantations. VREL's 567 full-time workers export 150,000 boxes annually via AgroFair. Women constitute 12 percent of the workforce. According to Gariba's fellow unionist Simon Adjei-Mensah, GAWU's "core activities are focused on the economic, social and environmental benefits of our membership and we make sure that plantation managers comply with ILO conventions and collective agreements. In partnership with IUF and ILO, we have developed specific activities targeting violence against women and child labour." GAWU is also involved in the per box social premium, "but it is managed directly by the workers ... to invest in business promotion, the development of social facilities, and in some environmental control." The union also organizes pineapple, cocoa, and mango plantations and has conducted campaigns around the WTO and EPAs (Adjei-Mensah 2007 EUROBAN-transcribed oral presentation).

In relying on FLO's commitment to basic Conventions of the ILO, Gariba Musah and Simon Adjei-Mensah can expect that they will not be discriminated against simply because they may be indigenous workers from a certain tribe (ILO Convention 111). They will not have to compete with conscripted labor (ILO Conventions 29, 105). They know their children under age fifteen will not have to work; those between fifteen and eighteen will not face hazardous work conditions and will be able to attend school (ILO Conventions 138, 182). Both will be guaranteed the official minimum wage (which many banana producers do not pay, despite national laws). According to FLO progress standards, within two years of certification, Gariba, Simon and their co-workers must receive a legal contract that has as its goal a living wage and benefits that include sick leave, overtime provisions, and a pension (ILO Conventions 100, 110, 111).

The combined power of FLO and ILO requirements has gained beneficial results! Gariba, Simon, and co-workers at Volta River Estates now receive a salary that is 60 percent above the minimum wage. At each growing area there is a health clinic staffed by a nurse. Management must guarantee GAWU workers safe equipment, clean water, gloves, boots, and a place to wash clothes used while applying chemicals (The aim of ILO Convention 155 is to prevent accidents and injury by minimizing hazards). Every FLO-certified plantation producer like the VREL must also provide a written commitment that workers like Gariba are free to join or organize an independent union which, once in place, must complete contract bargaining in less than a year (ILO Conventions 87, 98). If no union exists, workers are to form their own committee, which then would be given two years to achieve a mutually satisfactory arrangement with management.[20] General assemblies would meet at least once a year to approve their elected boards, staff, and annual reports. FLO created guidelines for monitors to verify how participatory the assemblies were, and the extent to which they encouraged an open process of planning, training, policy discussion, and decision making that enhanced member commitments.

To improve compliance with its requirements, FLO created FLO-CERT as a separate organization in 2002. FLO-CERT has had difficulty monitoring compliance with this last criterion—that workers had formed their own organization. It also faced problems in determining whether workers democratically reviewed proposed economic-development projects underwritten by the Fair Trade premium received from retail sales. Unions objected that FLO-CERT was accepting joint labor-management bodies, or "joint committees" as meeting this requirement, which would not assure true independence (see chap. 12). In 2005, FLO further revised its standards to enhance democratic elections and cooperation with local unions (FLO 2006).

Despite misgivings, COLSIBA hopes to replicate GAWU in Latin America, but it remains vigilant that ILO provisions be respected for obtaining collective-bargaining agreements. It asks that FLO-certified producers commit to consulting with national and international unions about worker representation, for example, the IUF. These commitments would represent a major departure from what banana companies have offered to workers in the past!

In the Meantime, an Insurance Policy

Despite the hope engendered by Fair Trade Social Development Standards, unions must primarily deal with nonfair-trade conditions. The COLSIBA network has mobilized to protect jobs in the banana sector and to negotiate worker-rights agreements with the major banana companies. In 1998, together with EUROBAN, COLSIBA organized the first major banana conference to discuss the global banana crisis. A transnational activist campaign with IUF support pressured Chiquita to discuss Ecuador's poor work conditions and low production costs with COLSIBA in 1998 and 1999. The TNC promised to respect collective contracts, and COLSIBA agreed to avoid actions that provided a competitive advantage to companies that violated labor rights. (Dole was cited for infractions in Ecuador, Fyffes in Belize, and Del Monte in Costa Rica [COLSIBA 1999, 2].) The Latin Union Coordination (or COLSIBA) requested Chiquita to take an historic step—a comprehensive regional worker-rights agreement.

Chiquita had already addressed certain social standards in working with the Rainforest Alliance. However, COLSIBA, EUROBAN, and USLEAP raised stringent objections that Rainforest certifiers did not properly understand labor issues. Rainforest Alliance leader Chris Wille ultimately acknowledged, "We can competently address conservation imperatives and some important worker safety and health issues, but when it comes to the emancipatory things like freedom of association, that's just not my bag" (Taylor and Scharlin 2004, 103).

In May 1999 the Costa Rican unions offered their own plan to Chiquita for free movement of organizers and elimination of rotations and blacklists and enforcement of laws that protect health and women.[21] COLSIBA and the IUF followed with a "summit" with all three major TNCs on market oversupply and low-wage labor. The companies recognized that labor rights were part of their future and agreed to set up a standing committee with the unions (which only Chiquita pursued).

In the aftermath of the 1998 hurricane, unions had lost some opportunities but gained others, broadened their network outreach, and expanded identity. In seeking to widen the definition of socially responsible banana production to include union rights, their efforts would have an impact on all certification schemes, including Fair Trade, in the way that Gariba Musah and his union in Ghana had come to assume and that COLSIBA yet hoped to achieve.

CHAPTER 9

Peasants of the Caribbean and Fairer Trade

WITH GOOD REASON, the image that most persuades banana alliance customers is the healthy farmer. Indigenous and peasant growers hold the sturdiest historical record for food self-sufficiency and earth-friendly cultivation despite some recent acquiescence to monocultural approaches to farming. In this chapter I briefly discuss the advantages of small-scale banana production. I subsequently trace the corporate and market pressures that eroded the smallholder approach in the Windward Islands. Fair Trade then stepped in with economic-development standards that benefited cooperatives. Farmers struggled over how to properly adopt FLO standards and gradually assume an identity as Fair Trade producers. As I point out in this chapter, Latin unions also came to appreciate the role of smallholders, even as they expressed reservations about the hiring of additional workers with little heed to labor rights.

Small-Producer Efficiency

TNC banana growers routinely point out that small-farm production is much less efficient than their own extensive banana-output system. Because Caribbean growers plant on divided, hilly plots with poorer soils that often lack irrigation, they harvest a third of the bananas that Latin plantations produce on similar acreage.[1] Peasants also find it difficult to cultivate bananas using the same uniform standards. While objective observers may judge the peasants' fruit delicious, TNC inspectors encounter flecks and bruises, or bananas that appear less tolerant of extended transport and retail shelf life. For them, holders do not offer a reliable supply of yellow-green produce that matches the demand cycle of corporate-controlled northern markets.[2]

Nevertheless, peasants constitute nine-tenths of all those who actually grow bananas. They view the crop as part of an ecologically balanced system that conveys nourishment, conserves forests, and protects habitats (Foro Emaús 2003). They cultivate banana shoots along with other crops that complement one another. While some easily achieve an organic rating, most minimize widespread pesticide applications.[3] Their expenditures on chemicals represent less than 10 percent of production costs, compared to 30 percent in monocultural plantation production (interview with Hermosilla 2004). Small growers avoid expenses associated with water contamination, pesticide-induced sicknesses, multiple plastic bags, waste disposal, and other monocultural damage, while gaining healthier soils. TNC and public officials do not routinely include these cost factors in their efficiency comparisons, yet the peasants' mixed-cropping model represents sensible sustainable-development policy for small island economies and other areas where other employment opportunities are few. Unlike the expansive production of sugar cane devoted to ethanol, integrated banana production reduces dependency because it contributes to food autonomy. It can stabilize populations and retard urbanization and shantytowns around tourist areas. Despite such persuasive reasons for pursuing this efficacious system of production, history demonstrates how the emphasis on uncontrolled, neoliberal, market mechanisms has forced a gradual weakening of the smallholder approach.

Smallholder Experiences in the Caribbean

As I explained in chapter 2, by early in the twentieth century, small Caribbean cultivators had already created a banana-production network that European colonial administrators and export companies sought to exploit. In the Caribbean, British Crown officials soon coordinated arrangements to market the fruit of thousands of tiny farmers in a development project that conveyed both benefits and drawbacks. Bananas became a major economic resource, garnering the Windward Islands dramatic foreign exchange and employment that continued into the twenty-first century. Bananas also represented an important wellspring of foreign exchange for the French colonies of Guadaloupe and Martinique. Jamaica and the Dominican Republic's dependence on banana shipments was lower than that of the French colonies, but Dominican exports grew in significance. Table 9.1 displays Caribbean national exports

TABLE 9.1. Caribbean Banana Exports by Weight (mt = metric tons) and Value (US$000), 1995–2006

Country		Percentage food/agri exports 2004	1995–99 avg.	2000–04 avg.	2005	Percentage change, 2000–04	2006	Percentage change from 2005
Belize	mt	30.6	51,754	52,357	64,891	0.24	72,699	0.12
	$	32.3	25,549	25,984	21,353	−0.18	25,296	0.18
Dominican Republic	mt	16.8	70,754	110,196	163,510	0.48	187,136	0.14
	$	8.9	12,311	30,800	44,640	0.45	130,910	1.93
Jamaica	mt	22.9	71,619	38,756	11,713	−0.70	31,863	1.72
	$	7.0	40,585	18,786	4,693	−0.75	14,783	2.15
Mexico	mt		184,465	59,158	70,166	0.19	66,599	−0.05
	$		59,732	16,710	25,342	0.52	33,351	0.32
Martinique/Guadaloupe[a]	mt		200,911	20,035	193,435	−0.03	172,433	−0.11
	$		137,639	113,076	153,419	0.36	113,538	−0.26
Suriname	mt	42.1	29,989	17,821	35,249	0.98	45,146	0.28
	$	38.9	18,234	18,728	13,996	−0.25	18,636	0.33
Dominica	mt	70.3	345,387	19,192	12,732	−0.34	12,852	0.01
	$	48.0	16,396	8,872	6,800	−0.23	7,490	0.10
Saint Lucia	mt	91.5	81,629	42,161	30,630	−0.27	34,935	0.14
	$	90.1	41,644	21,239	15,542	−0.27	17,761	0.14
Saint Vincent	mt	79.2	43,507	34,967	27,470	−0.21	23,783	−0.13
	$	86.9	20,153	15,016	12,815	−0.15	11,162	−0.13

Source: Computed from FAOSTAT.FAO.org (detailed trade statistics) 2008.
[a]Marketed through France.

by volume and price, 1995–2006. The first column indicates the percentage of national agricultural exports that banana shipments or revenues represented.

Banana cultivation in the West Indies must be viewed within the longer history of sugar production in the region. For many years, not only was sugar the primary export, it was also the vehicle for ongoing colonization (Wolf 1968; Bolland 1992). As the British government backed a plantation system that favored the settler class, it forbade land sales or transfers that might encourage competitive products or jeopardize the labor supply needed for sugar. Such a colonial agreement lasted until sugar production in Asia undercut Caribbean sales and employment in the late nineteenth century.

Given colonial domination over sugar, local small producers exported other products on their own. Although planting on the Caribbean's hilly slopes could be difficult, bananas could be successfully mixed with other crops. As sugar exports dwindled, small producers formed associations and started to ship bananas "as part of their diversified cropping system" (Striffler and Moberg 2003). When United Fruit spotted the trend in Jamaica, it quickly moved to control transport arrangements. Seeking independence from El Pulpo, in 1926 the British Imperial Economic Committee urged that its Windward colonies become the mother country's source for bananas (Grossman 2003, 290). Soon, the St. Lucia Banana Association, among others, was helping "farmers market their goods internationally through British, Canadian, and U.S. shipping companies," generating a 12 percent annual growth between 1925 and 1940 (Slocum 2003, 258).

Jamaica also notably increased its U.K. banana exports; however, its loss of shipments during World War II and climatic difficulties substantially hurt the industry. Fyffes (of United Fruit), the primary shipping agent for Jamaica, struggled for a resurgence in the 1950s, but the nation's low output dropped Fyffes's U.K. market share from 59 percent to 38 percent by 1969, as shipments rose from the Windward Islands. Jamaican producers claimed that United was undermining their industry by purchasing from elsewhere. After an official investigation, Fyffes/United agreed that, barring a shortage, it would import only Caribbean bananas.[4] However, island area hurricanes in 1979 and 1980 motivated Fyffes to invest in Belize and Suriname for additional supplies.[5]

Smallholder Arrangements in the Windward Islands

As Jamaica's difficulties persisted through the 1950s and 1960s, restive workers from the Windward Islands' dwindling sugar plantations inaugurated a series of strikes over pay and employment. Local colonial administrators, visualizing banana exports as an opportunity to minimize unrest (Trouillot 1988), offered farmers interest-free loans to buy and clear low-priced forest acreage for new plantings (Myers 2004, 20). While several area growers responded favorably, the primary banana impetus came from a new company: in 1954, Geest Industries, Ltd., negotiated a decade-long arrangement with four Windward Island Banana Growers Associations (St. Lucia, St. Vincent, Dominica, and Grenada) that it periodically renewed until 1995. Geest would ship all available fruit on a cost-plus basis, solidifying itself as the exclusive banana distributor for the associations in a way that redounded to Geest's benefit (Grossman 1998, 67). The TNC shipping firm also purchased several St. Lucia sugar plantations for its own banana production. By the early 1960s, Geest had strategically outmaneuvered Fyffes to gain half the U.K. market, with "green gold" dominating most Windward *and* smallholder output. "After Britain gave up its colonies, bananas became a way of life for the four tiny islands," explained Anton Bowman, a founder of the Windward Islands Farmers Association (WINFA). "This meant that if there was a change in that structure, there would be a change in this way of life"—signaling a change in farmer identity as well (interview 2005).

Transformations occurred both in Windward development and to the banana-production process. During the 1960s, Geest utilized the banana-growing associations to arrange marketing and shipping contracts (Crichlow 2003). The islands' smallholders took the opportunity to expand these associations. In comparison to the French islands of Guadeloupe and Martinique, where medium and large landholders achieved control, on Dominica and St. Lucia, associations assumed a major role in production. They offered crucial finance programs for fertilizer, local points of purchase, and broad access to markets. The associations actually bought the fruit and took on the risk, alleviating the pressure on small farmers. In the French Caribbean, in contrast, individual growers assumed all liability without assured sales (Welsh 1996).

Nevertheless, while Windward associations helped to protect mixed cropping, their autonomy declined as Geest's control increased, banana lands expanded, and the islands moved from being largely food self-sufficient to becoming food importers. By 1982, food constituted more than a quarter of imports (23.4 percent in Dominica, 27.5 percent in Grenada, 21.1 percent in St. Lucia and 29.2 percent in St. Vincent [Thompson 1987, 8]). By the 1990s, imports replaced more than half the local crops. "The banana industry served as one of the primary generators of capital and foreign exchange, and therefore, was given priority over other forms of agriculture," noted Susan Andreatta (1998b, 417). It hardly represented an impetus for sustainable development.

Major shifts likewise happened within the banana process. Although smallholders remained the more efficient producers, Geest saw them as an obstacle. The company cited competitive pressure from Dole and Chiquita, which were moving from stem to box shipping and applying stickers to emphasize brand quality (Myers 2004). Geest sought legislation, such as the 1967 Banana Growers' Act in St. Lucia, to suppress peasant influence. At the company's behest, the act specified rules the Banana Growers Association had to follow for fruit cultivation, shipment preparation, and handling. The act also enabled Geest to maneuver the association's conversion from a member organization to a statutory body. Local officials thereby became the monitors of production, harvesting, and processing practices in a way that wrought major changes in the lives of banana farmers (Slocum 2003, 262). Fruit growing had to be timed to meet Geest's shipping schedules. Workers had to rigorously follow procedures for planting, pruning, and applying nematicides, herbicides, and fertilizers, as well as for bagging the growing bunches to standardize "quality."

The changes had a major impact. The regulations exacerbated soil erosion and fertility loss. Geest advised farmers to apply more fertilizers, which over time increased land acidification and salinization. To kill weeds, the company also promoted a heavier use of paraquat and 2–4D, both of which retarded mixed cropping and species diversity (Andreatta 1998a,b). As they did elsewhere, field workers risked the hazards of farm chemicals despite their own knowledge and experience (Arcury, Quandt, Rao, and Russell 2001). Although ostensibly done to reduce work, the new procedures actually demanded greater labor input for less reward. Tapping British government grants and market access intended to stimulate smallholder output, Geest was transforming Windward farmers into a semiproletarian workforce.

The growers did resist, passing legislation in the late 1970s that forced Geest to lease or relinquish most of its plantation land. Yet, Crichlow (2003) argues that the company welcomed the opportunity to sell its property, letting governments and farmers assume production risks.[6] Island governments were persuaded to capitalize on traditional class and gender hierarchies of farming in which male heads of household put their wives and children to work on the land, and island states set up "model farms" that turned Geest's proletarian workers back into peasants. Nevertheless, as sole exporter, Geest still held the reins over local producers.

For a time, this system retained stability. The Lomé Convention of 1975 solidified special trade arrangements that proved very important for Windward banana sales (chap. 7). More than a quarter of U.K. consumption came from St. Lucia alone. Yet, this did not dissuade Geest from introducing a new cost-saving "field packing" scheme in the 1980s in which *farmers* had to pick and pack the bananas *directly* (Grossman 2003, 297). Such packing necessitated meticulously draining latex away from the cut banana hands and carefully encasing them with leaves in boxes to prevent bruising along bouncy roadways toward port. To meet "changing markets," Geest also required growers to sort bananas by size—all this for no additional pay. Yet the TNC persisted in earning significant margins from its guaranteed European revenues.[7]

"In the mid-1980s, farmers were called to produce more on the same land, with additional fertilizer and pesticides," elaborated WINFA's Anton Bowman. "This had already taken a great toll on the environment. Then came a demand for 'better cosmetic quality' or what we call the 'fairy prickle' banana. We had to use damaging pesticides and cover the fruit with nonbiodegradable plastic or the supermarkets would refuse the fruit. This added so much work, but we tried to cope despite rises in oil prices, which affected our input costs. At the same time the price for bananas was going down, which cut into what the farmer could pay the workers" (interview 2005).

By the early 1990s, green gold had become the preoccupation of Windward residents. On St. Lucia, more than a quarter of the workforce of 140,000 grew bananas on tiny plots averaging four acres, and up to three-quarters depended on the trade for their livelihood (Moberg

2005, 8). In Dominica banana production employed 60–80 percent of the working population; in St. Vincent, 54 percent. Under both state and Geest influence, the island farmers associations became increasingly market oriented, authoritarian, secretive, and class based, undermining smallholder identity and control (Crichlow 2003).

A Major Shift

In the early 1990s, the EU adopted banana-import regulations that weakened the Lomé agreement and eliminated specific national preferences, although it still guaranteed that ACP bananas would hold 19 percent of European banana imports (chap. 7). Officials touted "smallholder produce" as especially benefiting island women, who performed 60 percent of the required field tasks but received less than half the wages.[8] Yet the TNCs and large producers exploited the change to expand production to new ACP areas. Importers began buying less costly and more efficiently produced bananas from the Dominican Republic, Suriname, Belize, the Ivory Coast, and Cameroon, visibly angering Windward producers and officials (Slocum 2003, 264–65).[9]

The change in EU regulations also allowed importing companies to sell their licenses—to United Fruit/Chiquita's benefit: when ACP supplies were low, the license buyers could then bring in non-ACP bananas. For several years in the mid-1990s, weather conditions reduced Windward output so that island bananas reached only half of their quota.[10] Acquiring more EU licenses, Chiquita took the opportunity to increase its Latin sales. At the same time, the TNC lobbied the U.S. government to question EU guarantees for ACP bananas (see chap. 7)! Windward producers blamed the giant firm "on whose behalf the U.S. and Latin American governments challenged the tariff-quota system. . . . Chiquita's attempt to expand its sales by eliminating trade preferences for family farmers represents globalization in its most destructive guise" (Moberg 2005, 5).

Chiquita was not the only problem. Supermarkets complained that they could not get Windward bananas of sufficient quality, so they bought elsewhere. The new trade regime accelerated divisions between producers and the state-controlled grower associations. After 1992, the model farm program was dissolved in St. Lucia and reorganized elsewhere. Private purchasing groups emerged, and farmers felt forced to market their produce directly (Crichlow 2003).[11] Local holders bitterly reacted with a host of protests and strikes, especially in St. Lucia's Mabouya Valley (Raynolds 2003). Nevertheless, between 1992 and 1998 these factors precipitated a decline in Windward exports, for example, from 65 percent to 35 percent of the U.K. import total, adding fuel to the incendiary debate over banana-trade policies. In 1998, bananas still represented 40 percent of Windward exports, but the number of Windward growers had dropped from 22,000 to 15,000, and additional direct and indirect employment had fallen from 53,000 workers to less than 40,000. Exports between 2000 and 2004 were half of what they had been between 1995 and 1999 (table 9.1). By 2005, the numbers had dropped by yet another 50 percent.

While a notable number of Windward holders turned to marijuana cultivation as a defensive measure, island producers still sought ways to make their bananas more globally competitive. In 1995, they had set up their own shipping operation, the Windward Island Banana Development and Export Company (WIBDECO), which formed a consortium with Fyffes to buy out Geest and gain control of marketing (Myers 2004, 129). However, WIBDECO soon faced retailer and supermarket demands for sophisticated outputs and packaging similar to those that Geest had required. Some farmers were persuaded to experiment with new pesticide cocktails.[12] Others resisted the Banana Growers Associations' "need for attitude change in order to improve profits," such as keeping a strict regard for planting rules, field-packing procedures, and shipping timetables. The farmers decried the fact that 20 percent of producers already controlled 80 percent of exports (Lynn Allardyce, "The Impact of the Certified Growers Programme on the Windward Islands," (February 2000), cited in Liddell 2000, 9). Nevertheless, WIBDECO introduced the Certified Grower Program, which could trace product shipments and assure supermarket-quality bananas.[13] Its "Product Recovery" component sought to double yields in three islands but also to lower the number of growers from 8,000 to 6,500. Although the plan was designed to help exporters, several thousand small growers would lose out.

Such official policy appeared to render obsolete the Windward vision of strong, environmentally sensitive, farmer associations. After the St. Lucia Association imposed a banana quality-assessment and grading system that also eliminated "subquality" farmers, a 1998 survey found considerable pessimism and resignation among banana farmers. Half prepared to stop growing the fruit. Only those relying on unpaid family labor could make a go of it.[14] In St. Vincent, the Banana Growers Association had already introduced new methods for cleaning, labeling, boxing, and transporting bananas that further increased peasant labor and producer costs (Slocum 1996). Soon thereafter the association became privatized. The St. Vincent prime minister simply informed protesting smallholders that they had to face the reality of global markets.[15] Apparently, functionaries viewed Geest, Fyffes, and the European Union as benign agents that distributed Caribbean bananas within this much larger system; there was no other alternative (Slocum 2003, 267). By century's end, although bananas still represented an important percentage of Windward Island economies (St. Lucia, 62 percent; St. Vincent, 40 percent; Dominica, 23 percent), drug production had increased, networks were disrupted, and identities were compromised.[16] However, a fresh opportunity lay ahead.

Fair Trade in the Caribbean

Beginning in 1997, Fair Trade bananas arrived in the Caribbean, offering a direct-marketing alternative that promised sustainable development. While it supported quota sales and guaranteed access to European markets, it emphasized consumer awareness about choosing socially responsible bananas over choosing purely on the basis of price.[17] According to Lamb (2008, 29ff), a crucial British Cooperative chain's decision to feature Fair Trade proved essential for FTF endorsement and the revival of Windward banana production. In addition to criteria for environmental and social development discussed in chapter 3, FLO added economic-development standards that retailers were required to honor.

Economic-Development Standards

In offering more direct access to consumers, Fair Trade sought a fair price that assured commodity producers gainful employment on their own land. It assisted with product preparation and longer contracts. The global marketing process offered no such guarantees. But FLO even went a step further. To enable small farmers and workers to collectively determine their own needs, it required a "sustainable development investment" that buyers/traders remitted directly to producer associations. Banana retailers paid this premium of $1.75 a box to a special fund, referred to by VREL workers in Ghana. The premium was to be used, not for ordinary costs, but to enhance the organization's understanding of capital investment and risk management, or aid in meeting quality standards to assure export efficiency. FLO representatives often complemented this premium expenditure with instruction about market functions, environmental improvements, and technical training. Yet farmer associations could also expend premium funds on other community-identified priorities.

FLO made sure that producer associations decided democratically on how the premium would be utilized. Producer co-ops handled the fund a little differently than unionized workers but in both cases, a body of elected representatives was expected to make budget and spending decisions via a transparent process. According to FLO rules, the certified co-op, association, or union was to decide premium allocations in its yearly general assembly. If the majority of producer or worker representatives objected, a project would not go forward. FLO vigilance was in order since external funds could be maneuvered and cause enormous dissension. In 2005, FLO bolstered public accountability to "address the fear of management manipulation of worker organizations" (FLO 2006, 8).

Documentation of social-premium usage is beginning to accumulate. After his fieldwork in St. Lucia, banana scholar Mark Moberg found it the "most important contribution that Fair Trade groups make to their local communities" (2005, 12). Between July 2000 and April 2003, the Windwards gained nearly $1.3 million in premiums, with $750,000 going to producer groups. The rest helped cover WINFA and national committee expenses (WINFA 2003). As Moberg recounts, St. Lucia associations debated such premium projects at length, weighing the development contributions of mechanized "weedeaters" for replacing herbicides, the utility of school equipment, road upgrades, scholarships and occupational training for youth, and health reimbursements.

Nioka Abbott, of St. Vincent, whom we met earlier, believed bananas were "better than any other crops for regular harvesting. You get an income all through the year. That's why the banana is so popular as a cash crop. It would be hard to find a replacement." What Nioka found particularly attractive was "the social premium we get." The local farmers collectively decide its allocation, ranging from major projects like a health center to meeting immediate needs: "Last year we bought chairs," explained Nioka. "Before, when we held meetings everyone had to stand." Receiving the premium was just one reason why Nioka asked "people to buy more Fair Trade bananas and start putting pressure on supermarkets who don't buy Fair Trade. The market now is so small for Fair Trade that we need to get more supermarkets to buy them. If we could produce at a larger volume, then we'd get a larger income in return" (cited by Oxfam Great Britain 2004).

While not all associations functioned as smoothly as Nioka's, most experienced benefits. In St. Lucia, Moberg investigated six co-ops. The three with a collective history found it easier to meet their monthly meeting obligations, to disseminate technical and marketing news to members, and to notify higher levels about issues that required attention (Moberg 2005). However, nearly all the co-ops strained hard to meet FLO's social criteria that self-governing organizations were not to discriminate on the basis of age, gender, religion, or politics. This impressed a Jamaican investigator: "You have somebody who is illiterate but is able to sit in a meeting and contribute to a decision that will assist a particular school or build a playing field or basketball court. It is amazing to see the joy and sense of achievement among those poor farmers as a result of the social premium that Fairtrade [sic] offers them" (Abbott 2007). In the judgment of Senator Josephine Dublin-Prince, herself a member of the Windward Islands Farmers Association, "the Social Premium aspect of Fair Trade addresses and rewards care of the environment, good conditions of work, respect for human rights and justice. It is an excellent mechanism for promoting gender equity and equality" (2005, 92). The island's Fair Trade bananas are "produced under environmentally friendly conditions" by prohibiting most toxics and making "sure none leak into streams or rivers." The communities have depended on the Fair Trade premium "to invest in medicine for people with diabetes, buy computers and train the indigenous Carib people," explained WINFA leader Renwick Rose (T & G Publications 2004).

Some Difficulties

Despite their generally positive appraisal, Fair Trade premiums have encountered economic and environmental difficulties. Economically, although FLO-approved bananas garner a higher retail price, farmers do not always benefit. A test in U.K. supermarkets found FLO-approved bananas selling at double the conventional cost (a 100 percent markup), yet farmers received only a third of the increase.[18] FLO responded that it paid farmers on the basis of average Windward production expenses, although rates of pay on some islands (e.g., Dominica) exceeded the average. If prices from other ACP countries were notably lower than those from the Windwards, FLO might also lower the amount it pays.[19]

Equally problematic, Windward farmers are often assigned marketing quotas that restrict Fair Trade sales to less than half of their Fair Trade production. Even this half must be packaged in the field according to bewildering retail specifications reminiscent of Geest's requirements. For instance, "In August, 2003 . . . the St. Lucia Banana Corporation handled 14 different pack types for U.K. supermarkets" (Moberg 2005, 10).

Environmentally, FLO-approved producers face obstacles in meeting what appear to be northern-imposed regulations, as Moberg demonstrated. One example was the twenty-meter buffer required between roads and streams, quite appropriate for plantations but not for tiny hillside acreages. WINFA was finally allowed the right to plant fruit trees in the buffer. Another example was the substitutes FLO required for herbicides. Hand weeding is the most sensible approach, but unlike the unenforced minimum-wage conditions in Latin America, Windward growers had to pay relatively high fees of $15/day for additional labor. To avoid this cost, many utilized small "weedeater" machines that they obtained from the farmer associations. However, when applied to the indigenous watergrass, these machines broke the weed stems, causing rerooting that attracted more nematodes. From the growers' perspective, then, certain FLO rules exacerbated environmental problems. Both WINFA and CLAC farmers specified mandated pesticide substitutes as an unreasonable northern demand, along with pricing, supermarket packaging requirements, and fruit-inspection standards (Moberg 2005, 11).

Such complaints caused FLO to reevaluate how it approached environmental criteria. Although the Fair Trade project originated in the South, where smallholder ancestors understood how to grow chemical-free bananas, the imposed agrochemical usage by northerners like Geest had made it challenging to recover traditional practices. Jonathan Rosenthal insisted that FLO must vigilantly avoid another colonial scheme. While it did consult with individual producer organizations, FLO's board often did not represent southern voices or offer opportunities for organizations like WINFA to link together and formulate a common approach within FLO. In 2006, FLO's certification body inaugurated constitutional changes to bolster farmer input.

Despite these drawbacks, FLO's Economic Development Standards and certification brought benefits to the Windward Fair Trade network (as 2006 increases for St. Lucia and Dominica in table 9.1 show). Participants also internalized common understandings and cultivated a Fair Trade identity. Perhaps Oxfam, which promotes Fair Trade bananas, selected storybook examples when it featured several FLO-certified producers on its "Cool Planet" Web site (Oxfam 2004). Yet nuanced comments of George DeFreitas and Nioka Abbott conveyed more than poster propaganda. George DeFreitas explained that aside from a growing reliance on illicit drug cultivation,[20]

> St. Vincent is completely dependent on bananas. Whereas other crops might only be harvested once or twice a year, bananas give people a weekly income.... The benefits of Fair Trade for us have been reduction of chemicals on the environment. The social premium has also really made a difference. We're thinking about trying to open a nursery school and improve the roads around us. Anything we can do to make life better for the community, we'll try and do. We'd like to see the Fair Trade market increase as we have many farmers who want to become involved in Fair Trade. They too want a fair price for a fairly produced banana. (Oxfam 2004)

Nevertheless, explained George, "There are still some younger people on the farms but lots have left. Farmers paint such a gloomy picture of what it's like that the youth don't want to get involved. We depend heavily on being able to sell at a good price to a good market. If there was a growth in the market for bananas, then more people would get involved in production again" (Oxfam 2004) (fig. 9.1).

Regina Joseph began banana farming in Dominica when she was sixteen and has been a fruit grower ever since (fig. 9.2). In about 2002, when she was forty, she started with Fair Trade. Regina and her five children now harvest about seventeen boxes a week from her 2.5 acres, providing about 70 percent of family earnings: "I find that with Fair Trade bananas I get a better income." Regina spends the whole day at it. "Mostly women

FIGURE 9.1. George DeFreitas, Fair Trade farmer, St. Vincent. (Photo by Abigail Hadeed, Trinidad)

do the tasks like washing and packing the bananas. Sometimes my daughter helps me pack, but I always do the selection of bananas to sell myself" (Oxfam 2004).

Deryck, a youthful twenty-one-year-old banana farmer, worried that Fair Trade covered only a portion of his output: "I cut bananas on a fortnightly basis. My girlfriend washes and helps with the packing, and sometimes my dad might help out as well. The problem with the bananas you don't sell as Fair Trade is that the price always goes up and down. . . . My mum and dad have noticed a lot of differences over the years. They say things have got much harder now." However, Deryck expresses appreciation to people who buy Fair Trade. "Please keep on buying and then hopefully the quotas will get bigger and maybe we can sell to places like America and Canada. We've lost a lot here over the years. If things don't improve then banana farming will be a dead industry" (Oxfam 2004).

FIGURE 9.2. Regina Joseph, Fair Trade farmer, Dominica. (Photo by Abigail Hadeed, Trinidad)

By 2007, WINFA had networked two-thirds of island growers. Sixty percent of Windward bananas met Fair Trade criteria. The association stepped up its campaign for 100 percent participation, encouraging cultivation of additional crops such as coconuts as it matched production with new outlet opportunities. Following an FTF initiative entitled "Operation Perry," Britain's second largest supermarket chain, Sainsbury's, announced it would handle only Fair Trade bananas. To meet supply requirements for its 20 percent of the British market, it promised to buy 75–80 percent of St. Lucia and Dominica's output as a "quality investment" (Lamb 2008, 36ff, 155). "You have saved the banana farmers of St Lucia," the St. Lucia premier told Sainsbury's. "We now have young people who are coming into banana production who before could not be attracted," glowed Dominica's prime minister, who cited the social premium as an effective measure. "What better way to promote democracy in third world countries than allowing people to take on the leadership, management and the implementation of projects? . . . The islands of the

Eastern Caribbean have repositioned their bananas as a niche Fair Trade product. One can understand the tremendous positive impact that that has had on various communities" (Abbott 2007).

Organics and Fair Trade in the Dominican Republic

Although small farmers in the Dominican Republic began to emphasize banana exports only in the 1990s, by the mid-decade, the country had a fairly sophisticated system of organic production—exemplified by Heriberto Custodio and other Finca 6 families in Ázua mentioned in chapter 6. By 1999 the country had become the largest organic producer in the hemisphere, exporting 42,000 boxes/week (see Holderness, Sharrock, Frison, and Kairo 1999); by 2003, it had become the world's primary exporter, reaching 140,000 boxes, that is, 40,000 metric tons.[21] By 2006, the Dominican Republic exported 11 million boxes (200,000 metric tons) annually, with returns of more than $80 million (*Freshplaza*, September 18, 2006; table 9.1). Europe was the key market, with consumption totaling 140,000 tons/year—4 percent of total banana sales. British consumers had increased their purchases by 30 percent annually to more than £1.6 billion.

Dominican organic production has been a remarkable achievement. While the cost is 8 percent above that of conventional bananas, the fruit gains a 22–29 percent market premium for growers.[22] The Dominican model has focused on reintegration, "producing in 'natural processes,' encouraging trade . . . under certified organic conditions. . . . The organic movement currently goes further (than the fair trade movement) in revealing the ecological conditions of production" (Raynolds 2000).

Nevertheless, organic production in the Dominican Republic is not without drawbacks. Producer Christoph discussed changes in his company, Horizontes Orgánicos, since 2000. Horizontes

> works with fewer small farmers' associations than it used to. The entry into the market of larger organic growers (including Dole and Chiquita) is pushing out small growers by raising quality control and fruit appearance standards. Local cooperatives occasionally suffer from petty corruption and individual greed. Some farmers refuse to follow the guidelines

necessary for organic certification and lose their certification.[23] Others lie about the age of the bananas they are selling, which can ruin an entire shipment of bananas on its way to Europe or Japan, because one bunch of bananas ripening in a shipping container will release ethylene gas and cause premature ripening of the entire shipment. Strict production records must be kept and this is beyond the abilities of some of the smallest land owners. (cited in Boshart 2004)

According to the Dominican Organic Farmers Soil Association, by accelerating their market share, supermarkets have contributed to the trends Christoph described. "A significant number of small organic businesses have suffered from supermarkets switching suppliers or abandoning a brand in favor of their own label production," lamented Patrick Holden, soil association director. He complained that supermarkets were competitively pricing their organic bananas, offering lower prices to the farmers. "There is a tyranny about their own label products that allows supermarkets to abuse small producers." Even longtime farmers have difficulty surviving. Holden cited several suppliers whom supermarkets had threatened with delisting if they did not agree to sell directly or produce under the retail firm's label. The producers fear they might lose their business completely if they pursue the issue (*Freshplaza*, September 1 and 4, 2006, reporting quotes cited in the *Guardian*, September 1 and 4, 2006; see also Raynolds 2008).

Within Fair Trade, small farmers at Finca 6 have described similar pressures. Heriberto Custodio and his co-workers had formed the Association of Banana Producers in 1997 at the urging of the Fair Trade–certified exporter, Savid, which did not wish to deal with a plethora of independent farmers (Shreck 2002). Finca 6 member Angel admitted that in their prior experience, cooperatives did not control cheating, so they formed an owner association. "We want people, especially poor people, to be united. . . . If, like us, you don't [have money], then sometimes you're going to need a service from somebody that isn't money, a service given individually, personally between people. By being organized we can achieve this" (Ransom 2001, 92). The association was able to produce over two thousand boxes per week. Savid continued financing their output after Hurricane George hit in 1998. However, for the next several years, Savid retained the association's $1.75/box social premium to pay for a new road, irrigation equipment, and infrastructure recovery. Custodio still touted the premium and the association's benefits. The Finca 6 Association met regularly, elected officers who negotiated with Savid, and put pressure on any member who did not tend plants properly, since that would risk disease. It also decided on admitting new members. Custodio sometimes complained about having to put in extra work at the association to resolve disputes, but he found it much better than his former way of living.

Although social scientist Aimee Shreck concluded that the Finca 6 association began to make its own premium decisions only in 2000, after other debts were paid, she credited the experience with stimulating a glimmering sense of Fair Trade identity.[24] Yet given the low prices received at the time, most farmers she interviewed considered FLO an ordinary export program or an aid program, not a self-help participatory effort. The researcher faulted FLO for insufficiently educating Finca 6 holders about the significance of their participation. Only 51.3 percent said they benefited from membership; however, Fair Trade producers were 21 percent more likely to sell their fruit.

Finca 6 did confront new supermarket requirements. After 2000, with direct FLO guidance, the association took over selection and packing of bananas and distribution of payments to members. This created greater awareness, but it also meant producers had to stress fruit quality. They did so by organizing quality-control work brigades. For example, "Parceleros arrived to give Heriberto a hand. They worked with their machetes in an easy rhythm . . . chatting and joking as they went" (Ransom 2001, 93). Yet only the top fifty farmers were able to successfully produce Fair Trade output (Schreck 2002).[25]

Union Support

As Fair Trade expanded in the Caribbean, unions like COLSIBA endorsed FT initiatives that supported smallholders such as George DeFreitas, Nioka Abbott, Derrick Smart, and Heriberto Custodio. First, the unions believed it was the right moral choice. A fair price would keep local producers growing food on their own land in environmentally conscious ways. Second, unions knew that fair farmer livelihoods benefited labor's demand for similar treatment, and vice versa. When local holders won some control over marketing systems, workers also gained a voice.

Some employees would eventually become small producers themselves; for the others, heftier local control meant a stronger hand in negotiations. Unions likewise sought common ground with associations like WINFA, which they knew could supply only part of the market. "We have similar goals in making sure that the banana trading system protects both Caribbean and union bananas!" exclaimed German Zepeda of COLSIBA. "It is to our benefit to work as a strong alliance" (2005).

Nevertheless, unionized workers feared that when displayed side by side with unionized bananas, FLO-approved "smallholder" bananas were being promoted as "socially superior," when in fact they might not be. "Banana workers also deserve a fair wage," insisted Anton Bowman, a former St. Vincent unionist who later went into the banana business. "They work long hours and do difficult work" (interview with Bowman, 2005).

Summary

In this chapter I have tested peasants as potential actors in a banana alliance. I showed that smallholders can reach shared understandings and identities around trade fairness. Their interests can also mesh with the interests of trade unions such as COLSIBA and that of workers at the Volta River Estate in Ghana. I traced how European governments attempted to create a more protective trading system for local farmers that also benefited workers; nevertheless, smallholder associations also had to network, clarify their self-perception, and contest with state agencies and TNC marketing companies. When Windward smallholders replaced corporate-dominated transportation, distribution, and retailing systems with their own associations and premium decisions, they also began to appreciate the advantages of the Fair Trade approach. Yet a vision of equity for both small-scale producers and unionized worker networks faced internal and external obstacles. Fair Trade growers still confronted a corporate global system that jeopardized their livelihoods. Supermarkets emphasized "efficiencies" harmful to communities and nature, and FLO itself presented price and quota barriers. Northern "transnational activists" sought to reduce these obstacles with agreements on labor rights and certification programs, foci of the next two chapters.

CHAPTER 10

The Chiquita Accord and Labor Responses

IN JUNE 2001 Chiquita took a step that outdid the labor commitments of Fair Trade: it signed an historic worker-rights agreement with COLSIBA unions and the IUF, the first of its kind in the agricultural sector.[1] It committed itself to follow International Labor Organization conventions in all countries, and to ensure that its independent producers did so as well. It reaffirmed its previous declaration to address persistent worker health and environmental concerns, and it promised to engage with the unions when it considered relocating operations. Finally, along with the other signatories, it assented to a biannual monitoring process (see Taylor and Scharlin 2004, 152). The accord represented a major achievement in the fight for banana-worker rights! In conjunction with Chiquita's commitment to improve the environmental conditions with the Rainforest Alliance, it promised a vision for fairness that involved plantation workers, consumers, and a transnational company. Yet the accord soon revealed contending structural forces, network expansions, and identity considerations that advocates of Fair Trade and corporate responsibility had to reconcile. In this chapter, after an appraisal of the accord itself, I will explore implementation obstacles faced by both workers and Chiquita that have significance for all stakeholders in the Fair Trade debate.

Accord Basis

The notable labor accord that Chiquita signed in 2001 established the company's acceptance of worker-rights criteria that its unions had demanded (see chap. 8). Given United Fruit's many nefarious and bloody years as a major rights violator (poignantly described by Kepner and Soothill [1935]; Fallas [1975]; McCann [1976]; and others), labor leaders remained suspicious. After brazenly oppressing unions throughout its history, the fruit giant now inaugurated the new millennium by aspiring to a leadership position in worker relations. While Chiquita had admittedly become the most highly unionized TNC and alone had risked taking such a step, leaders and workers were nervous. With its stock prices at an all-time low, Chiquita was promoting corporate responsibility and labor cooperation as potential salvation, but how long would this last in an increasingly competitive market? In 2002, the firm even declared bankruptcy and reorganized.[2] For Ramón Barrantes, the leader encountered in chapter 1, "Chiquita signed the labor accord when its stock was $1. Now [by 2004] it is $20. True, the accord has given us some opportunities to pursue certain complaints. But it was a way out for them. Chiquita used the workers to help improve its financial condition yet the workers have received little in return" (interview with Barrantes 2004).

Still, the "heavily networked"[3] U.S. labor rights NGO USLEAP was elated. USLEAP had joined with its counterparts in EUROBAN to marshal a three-year campaign for the Global Framework Agreement. It represented "a significant step beyond codes of conduct . . . that envision little or no role for workers or their unions in monitoring compliance. . . . Rather [it] formalizes a conciliation process between the two sides." Yet, while the company had "moved far ahead of its main competitors in demonstrating a commitment to social responsibility," USLEAP warned that the real test would be how the agreement was implemented with independent producers in Ecuador and Guatemala (USLEAP 2001b, 6).

Union Appraisal

The accord presented local labor leaders with strategic and tactical dilemmas that Fair Traders would also face. Strategically, unions advocated member vigilance against the ever-present corporate tendency to weaken worker power; yet tactically, to protect worker rights, the unions had chosen to ally themselves with one of their worst historical enemies. Mistakes in resolving this dilemma risked undermining the achievement and identity of COLSIBA banana unions. "Ideologically, it was difficult to view Chiquita favorably," explained Costa Rican labor analyst Victor Hugo Quesada. "Why did the company destroy the unions in the first place? In addition, the Chiquita Accord did not say when free unionization could occur—it could be in two or twenty years. . . . And the bitter discourse from the eighties continues since real violations also continue. On the other hand, some [union leaders] don't wish conditions to improve because it bolsters their rhetoric of struggle, so this is a difficult and complicated problem" (interview with Quezada 2004).

Nevertheless, a review of the period between 2001 and 2005 found the unions generally content that the company had allowed open access to organizers and had set up a commission to resolve labor disputes.[4] Labor officials agreed that Chiquita was better than its competitors in honoring commitments. According to COLSIBA's Gilberth Bermúdez, "The Chiquita Accord has been a help to us. In Colombia, the agreement enabled the transfer of unions so that the new owners would respect free unionization. In Costa Rica, Chiquita became less antiunion than it had been in 1996–97. The accord helped resolve past cases in which it had supported Solidarista associations" (interview 2004). Chiquita's labor-rights strategy became a positive argument for maintaining European quotas (guaranteeing sales) and lower tariffs.

Worker Appraisals

More workers on the lush Atlantic coastal plain of Costa Rica produce the golden green fruit than anywhere except Ecuador. Bananas cover thousands of acres to the north toward Nicaragua and to the south toward Panama. They stretch by enticing palm-lined beaches below the port of Limón, then bump inland past the indigenous Bribrí region, and reach Sixaola, a narrow, muddy collection of cantinas and *tiendas* that straddles the double-lane gate at Costa Rican–Panamanian customs. Trucks porting bananas and other freight idle along the shoulders, their motors adding counterpoint to lively tunes that emanate from frontier cafes. On Saturday afternoons several kilometers back, employees gather in a concrete-block union hall surrounded by modest, tidy homes of banana workers to discuss difficulties (see fig. 10.1). Chiquita workers complain that neither the company's wages nor its environmental practices are up to standard. Jiménez Guerra, a Chiquita regular met in chapter 1, and Enrique, who works for a company contractor union, compare complaints. Florintino Chavez Calderon, the older worker with a wiry build who is the Chiquita union's secretary of organization, links gripes to banana sales: "The loss of our quotas will reduce wages, with big ramifications for the economy here," laments Florintino. "They are saying we can handle a tax of seventy-five euros a box, but if it is higher, it will hurt

FIGURE 10.1. Chiquita workers discuss global labor rights framework accord, Guatemala, 2001. (Photo by author)

us considerably. As a union, we are organizing to prevent this" (interview with Workers of Sixaola 2004; fig. 10.2).

Further north, in Sarapaquí, near the Nicaraguan border, workers voice parallel concerns as they gather for a weekend training session in the creaky but spacious second floor of a wooden-frame storefront. The union hall's windows overlook the main street of Villa Nueva, a provincial town more pleasant than Sixaola but just as hot. Most attendees are hired by Chiquita contractors; several work for Del Monte. All have yet to gain official union status. Among the participants, Marvin Matamoros Hernández, Abel Jarquin Gonzáles, Marcial Navarro, Theodore Sanchez Borge, and Marvin Zapata express similar views on pay and environmental safety, and add a few points about housing conditions and union rights.

Like Jiménez Guerra, from Sixaola, Marvin skillfully cuts the banana leaves that display sigatoka; as he explains: "[T]he leaf spores can infect other stalks. If three of a plant's eight leaves become infected, its fruit quality drastically declines" (interview with Workers of Sarapiquí 2004). Yet at 1,475 *colones* a hectare, Marvin is paid by the area he cleans, not by the stalks he prunes.[5] On ordinary days, he cuts three or four hectares,

FIGURE 10.2. Unionized Chiquita workers, Sixaola, Costa Rica: Raul Garcia, Union President; Jiménez Guerra; and others. (Photo by author)

earning about $2 more than Jiménez Guerra; however, if there is severe infestation, he is able to prune only two hectares, "not enough to cover my expenses." Abel works at Desarollo Agricola Industrial de Frutales, S.A. (formerly Caribana), which owns four fincas that sell to Chiquita. "The company assigns more land than it gives us credit for cutting. When it pays us 6,800 colones to clean four hectares, we are actually cleaning five" (Ibid.)

Labor rights are another crucial issue. "Our management does not want a union," insists Abel, as he named four workers he said were fired less than a year ago. "It invents reasons to terminate anyone who attempts to organize. Workers have filed many complaints with the Labor Ministry about violations of worker rights. However, the labor inspector is very corrupt and always declares that the workers do not have standing to file papers" (Ibid.).

"At our finca, the managers pretend that a free union exists," explain Marcial and Theodore. "Once the auditors are gone, this pretense disappears. Seven months ago they also began reducing our bonus that comes after working for six months. Some got no bonus at all. When we ask for our legal salaries, the company doesn't want to hear about it. They turn to a contractor who now is bringing in about 80 workers who are paid even less. Our union still has 20 members that are very strong within a workforce of 129, but the other workers are afraid about connecting with us" (interview with Workers of Sarapiquí 2004). In the view of these workers, Chiquita was not doing enough to make sure its managers and independent suppliers were complying with Global Accord provisions.

Leaders agreed with this assessment. Labor represented only 2 percent of Chiquita's operating costs, but as Barrantes and Zepeda pointed out, in addition to selective firings in Costa Rica, and backtracking in Guatemala, Honduras, Panama, and Colombia, Chiquita was buying more bananas from nonunion suppliers, costing even more unionized jobs. Bermúdez found it especially "hard to understand why Chiquita was transferring production to national producers on Guatemala's southern coast who are very antiunion when the company says it is committed to labor rights" (interview 2004). In 2008, Chiquita announced it would meet 30 percent of its European sales from Africa.

Chiquita Replies

Along with its union agreement in 2001, Chiquita received wide public praise for transparency in issuing a corporate social-responsibility report that not only acknowledged failures but specifically listed operations that were not in compliance (Taylor and Scharlin 2004, 153). The TNC subsequently published several annual updates, and then phased them out. But in response to problems raised by workers, Craig Stephen, president of Chiquita's Asia-Pacific division, proudly expressed that "for the first time in 100 years we have buried the hatchet and begun a process of dialogue. . . . Conflicts have remained; yet now we have better negotiations. For example on health and safety, there is no local agreement yet, but we are very happy with our progress" (Stephen 2005).

Enrique Vázquez, Chiquita's vice president for legal affairs in Costa Rica, acknowledged that the company still faced antiunion sentiments and must "work more on social standards," notably in the Costa Rica Sixaola and Bribrí regions, where Vázquez met with independent producers and the union "to resolve" atrocious wage violations—"paying one-third the minimum wage is unacceptable." The Chiquita vice president claimed that the company's efforts on behalf of national producers and contractors were another reason why it deserved certain tariff protection within the European market (see chap. 8). Vulnerability in dealing with associate suppliers affected Chiquita's direct investments: "We viewed Ecuador as a volcano—it could explode. We decided to leave Colombia because it was such a difficult place: Our workers feared being killed. So we elected to stay in Costa Rica, but here we face the highest production costs and it is difficult keeping some independents in business" (interview with Vázquez 2004).

Chiquita's Challenges

Although it had a profitable year in 2005, Chiquita faced increasing competition later in the decade. In 2006, the company experienced a 17 percent loss in some northern European markets and an 11 percent drop in its Middle Eastern sales. It skipped a quarterly dividend and raised its U.S. prices by 10 percent. It explored selling its Great White fleet (now down to twelve ships) and partnering with a specialized shipping company that could facilitate export of non-Chiquita fruit, including Fair Trade bananas.

Feeling threatened by intense market battles, after some management changes Chiquita also sought cost savings through work speedups and the closing of fincas. As a result, union attitudes toward Chiquita grew increasingly negative after 2005, especially in Honduras and Costa Rica. Workers and union leaders still praised Chiquita for taking a big step on labor rights and exposing itself to criticism due to imperfect fulfillment of the Global Framework Agreement; yet fresh failures risked scuttling the achievement.

Six Key Issues

Chiquita's accord violations reflected at least six cost-reduction efforts by TNCs and independent plantation producers engaged in race-to-the-bottom competition (table 10.1).[6]

TABLE 10.1. Banana Producer Race-to-the-Bottom Strategies

Strategy	Worker Impact
Shift production to national growers	Jeopardizes unions, more easily violates minimum wage laws
Increase use of labor contractors	Lowers wages, reduces benefits
Impose piecework pay arrangements	Requires additional work at same pay
Reduce traditional benefits (homes)	Raises living costs substantially
Minimally address women's concerns	Increases health risks, harassment
Minimally enforce ecology rules	Protects little despite appearances

National Producers and Unions

First, 75 percent of the banana trade still remains in the hands of four TNCs, yet labor leaders charge that they are purposefully reducing unionized output on their own land and *increasing* supplies from national producers. In recent years Chiquita has moved from a ratio of producing 60 percent on its own properties and 40 percent by independent growers, to a ratio of 33 percent from its own land and 67 percent from independent growers (Chiquita Brands 2007b).[7] By 2008 Dole was producing less than 29 percent on its own farms. This shift may parallel the TNC response in 2000, when oversupply conditions "allowed" fruit giants to cut purchases in Costa Rica but turn to independents elsewhere.[8] Or the TNCs could be maintaining the dual arrangement delineated by Foro Emaús leader Hernán Hermosilla: "In one [system], TNCs pay regular salaries of about $300/month; in the second, as in Sixaola near Panama, the independents pay $100/month to indigenous workers" (interview with Hermosilla 2004, verified by author). The TNC switch to the second system has become an increasingly obvious ploy to reduce labor costs. In 2007 in Guatemala, Ecuador, Colombia, and Nicaragua, Chiquita suppliers refused to negotiate contracts or pay social security (Zepeda 2007). In 2008, the company signed a long-term agreement with independent suppliers in Angola and Mozambique to supply 20–30 percent of its European market. When workers attempt to organize Chiquita-related producers, as they did at the Olga Maria plantation owned by Fernando Bolaños, Chiquita's biggest supplier on Guatemala's southern coast, they face more bloody lessons. In March 2008, assailants stormed the home of and gunned down Miguel Angel Ramírez, co-founder of the plantation workers' union (ACILS 2009), giving impetus to COLSIBA's call that trade fairness must also address corporate suppliers.

Contractors

Second, TNCs also hire temporary workers under various schemes. Some they contract directly under ninety-day provisional arrangements. After a brief hiatus the TNCs can either rehire the workers, or sign them indirectly through a subcontractor. Migrant workers like the half-million Nicaraguans residing in Costa Rica are especially vulnerable to such exploitation. Union officer Auria Vargas Castañeda believes subcontractors bring in workers from Nicaragua and Panama "to fix the cables, but then abuse and steal from their employees" (interview with Vargas Castañeda 2004). Social researcher Maria Eugenia Trejos finds subcontracting becoming common in Honduras: "The companies exploit these workers. Their primary approach is to fire and recontract regulars at reduced salaries." Trejos and her team surveyed a Del Monte/Bandeco area in Costa Rica where only 20 percent of the workers were permanent. They discovered "temporary workers contracted by the company directly, those paid by a subcontractor by means of a daily pay stub, and even some workers contracted for a certain task/area who themselves arrange for others to help them" (interview with Trejos 2004). Human Rights Watch (2002) and Banana Link (2000) documented similar hiring complexities in Ecuador.

Despite improvements, contract employees still face austere conditions. In unannounced visits, geographer Dr. Carrie McCracken found that contracted workers in Costa Rica "often work very long days, making less than minimum wage. Yet because of NGO attention, companies like Chiquita, Dole, and Del Monte have been forced to clean up their treatment of day laborers." The geographer estimated that the TNCs were using fewer subcontractors by mid-decade and almost all employees on three-month contracts had health benefits (interview with McCracken 2004).[9] Chiquita's Enrique Vázquez explained how a policy he helped design had inaugurated a fairer approach since it allowed contracting under only three conditions:

1. Necessary temporary work specifically identified, such as cleaning drainage ditches. Workers could be hired every eight months for three- to four-week periods;
2. Fertilizing (although this was usually assigned to permanent workers);
3. Applying agrochemicals when the workers are hired by the furnishing chemical company, since they are trained to spray nematicides and herbicides without problems.

Chiquita's policy makes sure that contractors pay social security and worker compensation in case of injury. The TNC retains 5–16 percent of the agreed fee "in case the contractors miss making such payments, which they often do" (interview with Vázquez 2004). Vázquez acknowledged that despite this policy, a lot of producers use contractors who exploit undocumented Nicaraguans and Panamanians. However, Chiquita has "helped many with migratory papers. We think it is better to have them in the permanent workforce even if they return to their country. We have also reduced ninety-day rotations to a minimum" (Ibid.).[10]

Despite Chiquita's intentions, all workers at COBOL in Costa Rica were on six-month contracts in 2008. Findings showed most contracted workers were not receiving health benefits (Harari 2005b). Subcontractors forced workers to surrender 9 percent of their salary, supposedly for social security, which the subcontractors routinely pocket without paying the 27 percent employer obligation. Contracted workers also gain no vacation or pension or the thirteenth-month bonus (*alguinaldo*) that regular employees receive (interview with Hermosilla 2004). For 60 percent of the banana TNC workforce, the twenty-first century contracting reality appeared very different from Chiquita's stated policy.

New Work Arrangements

Third, competitiveness stimulates firms to extract more work by intensifying assignments or introducing new technologies and incentives. Every union that responded to a recent survey cited how companies squeeze workers by adding extra tasks for the same pay, or they convert to a piece-rate system for remunerating bunches packed or hectares cleaned.[11] In

Honduras and Guatemala, Chiquita implemented *caja integral*, a new method for compensating workers based on the amount of exported product rather than on the tally of picked fruit. In Honduras and Sixaola, Costa Rica, Chiquita removed cable motors that pulled fruit to packing sheds, forcing workers to hand-carry the huge stems. Leader René Garcia Miranda saw a contract violation: "We have been using these motors for twenty-three years!" René wondered why a company emphasizing quality would return to backbreaking human labor. "At least it could investigate a mule transport system such as the one used by Dole" (interview with Garcia Miranda 2004).[12] "Companies are creating mega-fincas in which work teams are assigned more tasks, additional fertilizing, etc. The number of boxes shipped increases, while employment remains at the same level. It all adds up to more exploitation" (interview with Hermosilla 2004).

After 2002, when Chiquita encountered resistance to new work arrangements in Bocas del Toro, Panama, it substantially reduced purchases, and 830 workers missed a payroll. Chiquita resumed regular purchases in September 2006 but still demanded caja integral to calculate worker pay, stimulating protests over contract violations.[13] At Armuelles, on the Pacific, 2,800 COOPSEMUPAR members faced unemployment and bankruptcy because of "a corrupt transitional contract." To retaliate, COOPSEMUPAR rejected a $10 million investment agreement by Chiquita and the government.[14] To protest Chiquita's low process, SITRACHILCO and COOPSEMUPAR distributed fifty thousand boxes of free bananas to the public.

Chiquita has acknowledged that banana production requires long, hard days with too many hours. "Yet," says Vázquez, "some workers prefer days of ten hours so that they can earn more" (interview 2004). However in late 2005, when Hurricane Gamma brought more flooding inundating Honduran farms, the company decreed that it would close Tibombo, where SITRATERCO had a long history of militancy, and Buenos Amigos, a freshly organized finca. Honduran officials lambasted the company for terminating two unionized plantations and implementing "draconian productivity demands" on others. Negotiations ceased when the company proposed raises "based on production standards impossible to fulfill" and threatened workers that if they wouldn't accept, the firm would abandon production (Zepeda 2007). Although Chiquita had offered early retirement to hundreds of older Buenos Amigos employees, it forced speedups on the remaining workers, so the union claimed a violation of the Global Framework Accord. After a bitter exchange, and intervention by the IUF and COLSIBA, Chiquita agreed that Buenos Amigos would become an independent supplier that recognized worker rights and union affiliation.[15] However, the new owner refused to recognize the union, causing workers to lose benefits and protections. Zepeda questioned how Chiquita could take such action while it touted its social compliance in Europe under Rainforest and SA8000 auspices.

Benefits Elimination

Fourth, national firms and TNCs are severely cutting traditional worker services. They are "reluctant to reclaim the high level of social responsibility that became increasingly burdensome as the high-profit years of paternalistic enclave agriculture came to an end," explain Taylor and Scharlin. "All have come to recognize that the previous kind of worker dependency is not sustainable" (2004, 115). In all countries but Colombia and Panama, transnational firms attempted to sell their housing to the workers or let it fall into disrepair.

Guatemalan unions claimed this violates their contract. Rather than paternalism, they view housing as a long-fought-for right, and they resent company efforts to unload dilapidated units that TNCs had promised to maintain. In Sarapiquí, Costa Rica, employees challenged the removal of housing the company had traditionally furnished. According to Marvin Zapata, at the El Fin plantation, "The company wants to sell the housing to us, but most of it is in bad condition. The walls and doors are falling down and it lacks water; yet much of it is thirty meters from the river and susceptible to flooding. They say we have to pay $1000 or leave the plantation" (interviews with Workers of Sarapiquí 2004).

"In our agreement, the company has an obligation to provide housing," chimed in Marcial and Theodore. "It is our patrimony, and the union has been fighting to protect it. They could build a sewage system to remove the water from houses in low-drainage areas, and provide bridges to access the main road" (interviews with Workers of Sarapiquí 2004).

Besides housing, other benefits are in jeopardy. Companies have reduced staff at medical clinics. In a 2005 survey, the banana unions agreed that although the laws prescribe it, few workers take a pension. TNCs are also abandoning support for finca schools, which six unions rated as notably deteriorated. In Ecuador, employers do not provide latrines and potable water (Harari 2005b). In Sixaola, Costa Rica, Chiquita even offered to sell its land to the workers. "Our assembly said no," explained union Secretary General René Garcia Miranda. "The company could walk away, and we would lose our pensions, and the benefits of the current contract, besides having to pay all the land transfer costs" (interview 2004).

Women Workers

Fifth, banana unions have increasingly demanded more respect for women, who constitute the majority of packing-plant employees, and total about 100,000 in the sector worldwide. Almost 60 percent are single-mother breadwinners and 35 percent are illiterate. Despite company commitments, the women feel they have to struggle disproportionately for pregnancy leave, nursing facilities, and child care remedies, as well as resolving sexual harassment and wage-equity problems (Munguía 2005, 88). According to Banana Link, women earn only a quarter to a third of what nonunion male packers make. In addition, "they are more vulnerable when their piece rate work is evaluated by male supervisors" (Longley 2005).

Health remains another major preoccupation. Even though companies have installed protective spray boxes, in almost all packing plants the women still brush fungicides by hand to prevent banana-stem crown rot. McCracken reported many instances of "a woman with a brush in one hand and a cup of fungicide in the other. Painting the mix causes a lot of arm burns. Most women now wear gloves; and some plants have installed a tube that feeds fungicide to the brush. Yet even women who paste on the labels inhale the fungicide. It gets on their clothes and they suffer a higher rate of sterility" (interview 2004; see also chap. 6).

One of Auria Casteñeda's tasks as women's secretary (and financial secretary), is to "keep constant vigilance" to prevent health problems and sexual mistreatment.

> By contract, when a woman is pregnant, she must be assigned a safe job without chemicals, and for this I need to take a firm stand. It takes some

doing to convince a supervisor. I have to say, "Who will take responsibility for the child, you?" The law also provides that a pregnant woman be given paid time to be with her baby, one month before birth and three months afterwards. She must be given an opportunity to feed her baby for another two to three months while working in the plant. Because of health issues, she cannot feed the baby inside the plant; so I make sure the company allows her to return to her home at 10:30 a.m. and return at 1:00 p.m. . . . The Chiquita labor accord helps assure company compliance. Now I can raise the regulation for transferring pregnant women to a safer environment.[16] (interview with Vargas Casteñeda 2004)

Yet to avoid liability, companies have substantially decreased their employment of women. "We are faced with the worst options," insisted Iris Munguía, COLSIBA's secretary for women.

The enforcement of our maternity and breastfeeding leaves make our social cost more expensive. For instance, in Honduras, up until 1998, 2,500 women worked in the farms, whereas only 1,000 of us are working there now. In 1995 in Colombia there were 1,700 of us. Now there are only 1,250. In Costa Rica, between 1998 and 1999, women accounted for 35 percent of the labor in packaging sections. This percentage has now dropped to 25 percent, according to the employers. However, banana union leaders argue that these percentages are actually much lower. Since the implementation of the Equal Opportunities and Maternity Act, employers have reduced contracts with women as a repressive measure. (Munguía 2005:87)

In their study of Chiquita, Taylor and Scharlin acknowledge that the company should be doing more for women (2004).

Despite insufficient TNC engagement, banana women are taking union action. In 1985 they grasped the opportunity to form their own section, which has continued to expand. In 2000, via a COLSIBA/ASEPROLA (Asociación Servicios de Promoción Laboral [Labor Promotion Service Association]) project, thirty-five women workers compiled life histories documenting their issues. Two years later, COLSIBA held gender workshops that trained women in ways to participate in union and NGO organizations while meeting practical family needs (see http://www.colsiba.org). In 2004, twelve Central American activists collaborated with the textile support group STITCH in Guatemala to design the "Women, Labor and Leadership Training Curriculum." According to STITCH director Beth Myers, the program encouraged "women to take leadership roles in the union and to press for adequate maternity leave, prenatal care and freedom from sexual harassment in all contract negotiations." Over two hundred women on Del Monte plantations in Guatemala, and additional women in Honduran banana labor federations, have received the training. For example, Iris Munguía conducts "workshops on gender issues with groups of both men and women to raise awareness." The curriculum also teaches women about "the financial value of their contribution to the household" and "challenges the assumption that women working outside the home will also carry the load of domestic work and the gender-based separation of jobs in the banana plantation." Currently, "only men move boxes into shipping containers and only women do the lowest-paying jobs, such as sticking on brand-name labels." As a result of these training sessions, the Honduran union has set up women's subcommittees on twenty-five plantations (Wisniewski 2006). In *Bananeras*, Dana Frank wonderfully documents additional involvements (2005).

Environmental Policies

Finally, while the companies have made environmental improvements (chap. 6), workers experience spotty compliance. Despite Chiquita's rule that employees be kept out of newly sprayed fields for thirty-six hours, Chiquita's supplier Caribana "sprays continually when the workers are in the fields, creating skin lesions that causes scratching. The company does not provide sufficient protective equipment such as gloves and vests. It also has no washing facilities in the plant, so we wash clothes at home, but this contaminates our other clothes" (Gonzáles, interviews with Workers of Saripiquí [2004]). Marcial Navarro and Theodore Sanchez Borge, from the plantation Gacela, verified that aircraft spray the fields routinely. "Although our management says it is complying with standards such as SA8000 and ISO 14001, I wish I had a camera!" exclaimed Marcial. "When the auditors visit, they send those of us loyal to the union to work in the far cable area or switch us to another finca, so we won't be available for questioning" (interviews with Workers of Saripiquí 2004). Florintino Chavez Calderon described how the spray "once came down in a liquid and burned my arm." He then took a job boxing bananas in the plant. Yet when planes fly over the fields now, the company generally gives "warnings to keep people away" (interviews with Workers of Sixaola 2004).

Nevertheless, in 2007 Alexander Reyes Zuniga, Jaime Blanco Juarez, and Marco Gonzalez Borge told Chiquita/Cobal supervisors in Costa Rica that toxic nematicide spraying had made them sick in the area where they were harvesting. Blanco visited the doctor for tests. Reyes remained nauseated the next day but was ordered to work. When the team complained, supervisors accused them of entering the doused area contrary to orders. They fired Reyes and Borge for misconduct without any further examination or required warnings (Banana Link 2007). After the local union failed to gain reinstatements, COLSIBA enlisted its transnational activist network. BanaFair's careful on-site investigation demonstrated how the incident undercut an environmental award Chiquita had just received, showing "a completely different face . . . than the one which is projected throughout certifications for 'sustainable agriculture practise'" (Fischer 2007a). "Chiquita is becoming even more anti-union and hiding behind its certification from RA, SAI, ISO, and the ETI Initiative" (Fischer 2007b). Publicity by Labourstart/U.K. rapidly generated appeals from 3F, Peuples Solidaire, Banana Link, USLEAP, and others. Finally, the company acknowledged local management error and reinstated the workers (USLEAP 2007).

Union Freedom

Underlying the six race-to-the-bottom strategies of utilizing national producers, subcontracting, reorganizing work, denying benefits, mistreating women, and environmental backsliding is the denial of union freedom. Costa Rica remains archetypical. In the mid-1980s, as I recounted in chapter 2, banana TNCs decimated unions and replaced them with employer-controlled Solidarista associations. Only gradually was the COLSIBA network able to inspire new organizing. "We thought the Del Monte accord [in 1997] would be an opportunity for growth," explained social scientist Maria Eugenia Trejos, "but then came the oversupply crisis, firings, and repression, and the unions were unable to recruit new affiliates" (interview with Trejos 2004).

In 2007 COLSIBA publicly asked its transnational network to contact the company. Despite the 2001 accord, Chiquita terminated union members and threatened to close its Sixaola plantations in Costa Rica. When labor leaders suggested a remedial program, management laid them off. COLSIBA likewise cited productivity demands and bargaining failures in Guatemala and Honduras, and the lack of contracts at Nicaraguan suppliers as evidence of Chiquita's accord commitment failures. Violations were creating a "new, tense stage in relations between Chiquita and its unions" (USLEAP 2007, 3). EUROBAN lamented Chiquita's firing of union officers, contrary to ILO Convention 135, and its inability to hold suppliers "Banacol and Desarrollo Agroindustrial de Frutales SA/Caribana" accountable for accord violations.[17] The TNC reportedly would not challenge suppliers that promoted Solidarista associations; instead employers brazenly argued that the ILO reevaluate their merits. Back at the table, Chiquita, COLSIBA, and the IUF formed mechanisms to alleviate disputes in specific countries.

Chiquita was not alone in backtracking on union rights. Despite dialogue in 2007, Del Monte and Dole demonstrated very limited willingness to consider any global accord. Marco Garcia, head of Del Monte Guatemalan operations, recognized the inhumanity of "workers pulling in fruit by hand," and boasted that since 2002, Del Monte had stressed "labor-management relations as an example of better labor conditions . . . (and) signed collective agreements in record time . . . Collective agreements are necessary, and both parties must respect them" (2005). Nonetheless, Del Monte subverted contracts in Costa Rica and persisted widely in sourcing nonunion bananas. It formed a joint IUF commission to improve SITRABI worker security following the 2007 murder of Marco Tulio Ramirez but wrangled over costs. Dole, the least communicative with the unions among the big three TNCs, claimed a rigid verification process for free association embodied in the SA8000 code (chap. 11). Yet despite a 2007 settlement in Costa Rica, Dole retained a poorer labor-rights record than Chiquita (see Dole Campaign, chap. 13).

All companies and countries promoted an image of social leadership as they squeezed labor costs and undercut union strength. "The monitoring programs claim there is free unionization. They have certified 80 percent of Costa Rican fincas; but where is the union freedom?" demanded Gilberth Bermúdez. "We had to file some thirty ILO complaints, eight in the last five years" (interview 2004). Altogether, between 1995 and 2005 more than ten thousand unionized workers were laid off as a result of closures, large-scale dismissals, and disease problems (Zepeda 2005). Unions reported their affiliate numbers had been cut by nearly half (Harari 2005b). "Attempts to organize workers provoked a shift of production . . . such as those to the southern coast in Guatemala where workers labor as much as 14 to 16 hour days, with no overtime pay, no extra compensation for holidays and Sundays, and are obliged to get up at 3 a.m. and return home at 10 p.m. (interview with Zepeda 2006).[18] In the workers' view, without strong unions, none of these issues would be adequately addressed.

Questions of Union Identity

The COLSIBA network functions a little like the United Nations, not like the TNCs: its member organizations hold most of the power, not the network itself. While COLSIBA affiliates share vital interests, they also reflect differing national experiences. Their hesitant reactions to the Chiquita Labor Accord illustrate COLSIBA's challenge as it seeks a shared sense of "who we are" among unionists facing changing structural conditions.

One affiliate that had difficulty developing a common identity was UNSITRAGUA in Guatemala. Owing to the TNC's reorganizing initiatives in 2004, the union faced the closure of half of its Atlantic coast affiliates. Meanwhile in Honduras, Chiquita implemented caja integral, the new method for paying workers based on the amount of exported product. Union leaders realized this would mean working for less pay; nevertheless, since they could negotiate the terms for how caja integral might be recalculated under the Global Accord, the affiliates could also minimize finca closures. Taking this approach, SITRATERCO in Honduras gained an agreement for computing caja integral that also prevented most finca terminations. In Guatemala, in contrast, UNSITRAGUA did not identify the rationale behind the global accord. Taking a hard line toward Chiquita, it refused to discuss most work-rule changes or accept any closure agreements. Although Honduran union officials offered to assist, the Guatemalan representatives rejected what they considered a compromising strategy. As a result, Chiquita terminated half of UNSITRAGUA-controlled fincas. Knowledgeable observers thought some could have been saved, but since UNSITRAGUA negotiators clung to dogmatic notions about class enemies, the organization could not formulate a common strategic identity. This prevented the membership from exploiting their limited capabilities within the global accord. One can sympathize with UNSITRAGUA's position, given how Chiquita's later impositions on Honduran and Costa Rican unions called into question the accord's functionality. Although UNSITRAGUA fincas gained a 15 percent raise in 2008, earlier rigidity illustrated the challenges COLSIBA faced as it mobilized to take TNC action. Any mistakes COLSIBA made in resolving affiliate dilemmas risked undermining its achievement. TNCs such as Chiquita could quickly exploit eruptions within COLSIBA and between COLSIBA and the IUF. While disputes within democratic networks can be expected, they illustrate obstacles that unions must overcome to maintain unity in a period of global competition.

Conclusion

In the first decade of the twenty-first century, workers appeared more open to the possibility that some of the companies for which they worked and the countries where they lived were addressing certain labor concerns. The Global Labor Accord with Chiquita brought some benefits to banana workers. In its efforts to address social conditions, the company demonstrated leadership beyond that of other TNCs. But all TNCs, Chiquita included, continued skirting commitments by shifting operations to national producers, subcontracting, reorganizing work, eliminating benefits, reducing female employees, carelessly spraying, promoting Solidaristas, and avoiding unions wherever possible. Worker experiences with Chiquita illustrate the difficulties unions face in gaining representation. In their debates over acceptance and implementation of the accord, some COLSIBA affiliates missed opportunities, clung to dogmatic attitudes, and failed to appreciate changing competitive conditions. This jeopardized forming a common sense of unity that Chiquita and others were able to exploit. It also reduced labor's identity in forming a coherent engagement with Fair Trade's social-development criteria. Nevertheless, the lessons drew attention to labor rights as a fairness issue. They helped prepare producers such as WINFA and COLSIBA affiliates, and NGOs such as TransfairUSA, Oké, AgroFair, and Oxfam to mindfully address other dilemmas of the banana system, as I discuss in the chapters ahead.

CHAPTER 11

The New Banana-Marketing Strategies

AS PART OF their own ethical identity, northern consumers have focused on employment conditions in the South. We earlier saw how the Rio Earth Summit and subsequent international gatherings drew public attention to environmental issues. Union networks exposed pesticide contaminations, worker sterilizations, and violent repression. Foro Emaús, Banana Link, and EUROBAN promulgated magazines, videos, and Web sites that documented fish kills, firings, and deaths, whetting consumer worry. As shareholders and purchasers understood *how* bananas were being grown and sold, they questioned retailers and pressured TNCs. Although the World Trade Organization (WTO) and regional trade bodies had largely dismissed social and environmental concerns, banana marketers discerned that an emphasis on banana "quality" helped assure a skeptical public.

In this chapter I examine third-party corporate certification as a fresh alliance opportunity. Although major firms such as Chiquita and Fyffes made early attempts to design internal codes that addressed employee and health conditions, the public quickly dismissed any company's ability to audit itself (Esbanshade 2004; Chambron 2005). During the 1990s, however; worker, customer, and retailer interest precipitated a plethora of "independent" third-party monitors that certified bananas. Programs became more adept at evaluating ecological improvements than auditing social conditions, since the latter could prove costly. Yet the proliferation of all programs added expense and confusion. The various monitoring approaches enabled banana stakeholders to stress their environmental and labor practices as a selling point in "open market" conditions. NGOs and unions asserted that companies and retailers exploited the auditing process while undercutting collective bargaining for competitive advantage. After deciphering whether TNC certification contributes to a responsive consumer "identity," I outline four difficulties unions and smallholders experience with codes, and how these difficulties may jeopardize a producer-consumer alliance. I show that product certification can be either an opportunity for building networks and expanding public awareness about the arduous aspects of banana production or an unfair promotional gimmick that deceives consumers and obfuscates exploitation. In chapter 12 I further pursue Fair Trade approbation as an alternative to corporate certification.

TNC Programs

Beginning in 1992, as I noted earlier, Chiquita collaborated with the nonprofit Rainforest Alliance to improve the environmental conditions of banana production via the third-party Sustainable Agriculture Network (SAN). SAN set targets and calculated compliance for reforestation, chemical use, and water quality. By 2001, Rainforest/SAN had certified all Chiquita plantations (Taylor and Scharlin 2004). In the mid-1990s, Dole and then Del Monte adopted the ISO 14001 environmental-management system, which, unlike the SAN, gave companies considerable latitude in creating their own benchmarks. Auditors such as the nonprofit Bureau Veritas Quality International (BVQI) and Société Gènérale de Surveillance (SGS), or the for-profit Ernst and Young, specified how ISO procedures should be carried out but did not specifically state what the firms needed to address. Since neither SAN nor ISO 14001 devoted much attention to the social aspects of the banana trade, other programs emerged that did. Via official U.K. auspices, companies and NGOs created the Ethical Trade Initiative (ETI), which established baseline social standards. From the United States came Social Accountability International and its SA8000 checklist for referencing core ILO conventions. As the European trade regime turned away from designated banana suppliers, banana TNCs sought to capture market interest and emphasize their environmental and labor practices by expanding third-party certifications listed in table 11.1.

SAN

The Rainforest Alliance's Sustainable Agriculture Network program stresses conservation, better workforce treatment, and improved management systems, according to Technical Manager Tom Divney. By 2005 annual SAN audits had certified 243 banana operations in nine Latin

TABLE 11.1. Banana Third-Party Certification Programs

Program	Environmental Standards	Social Standards
Rainforest/SAN	Rigorous benchmarks	Limited rights, audits
ISO 14000	Management system only	None (see below)
ETI	Commitment, but no audits	ILO rights, no audits
SA 8000	None	ILO rights, rigorous audits
Eurepgap	Phytosanitary standards	Worker health only
IFOAM organic	Rigorous benchmarks	Unenforced recommendations
Fair Trade	Rigorous benchmarks	ILO rights, audits
Other	Benchmarks	Limited rights, audits

American countries and the Philippines covering 146,000 acres. It had also verified most of Chiquita's independent producers. Rainforest-approved bananas shipped from Latin America totaled 1.3 million tons, 15 percent of the region's total shipments (FAO 2004).

Rainforest prided itself on "using local auditors with local knowledge and culture who also provide lower costs for producers," shifting the expense "to the brokers and retailers." In collaboration with the U.S. Environmental Protection Agency, it used Integrated Pesticide Management to "drive down the use of pesticides, especially for sigatoka." Divney hoped to open markets to farmers who want to do things better. He aspired to educate retailers to seek certified products. "People know nothing about banana production. It is scary in the fields, with more piece rate contracts, no overtime and outsourcing without responsibility.... Corporate supplier codes help business to do things correctly." However, Divney acknowledged that "audits are like a photo in time. Certification does not replace a union. We like to think we open the door for unions and have a good relationship with the IUF. I tell producers, do a good job with the people; they will produce for you, and this is the true path to economic sustainability" (Divney 2005).

According to the Swedish Society for Nature Conservation, Rainforest has the best criteria for protecting biodiversity and effective measures for preventing soil erosion and handling waste (Mattsson 2004). The FAO determined that the price for implementing SAN in Costa Rica added only 1.5 percent to production costs, allegedly recuperated by retailers who advertised certified bananas. In mid-decade, by combining the Rainforest "Frog" logo with its own Carmen Miranda label, Chiquita was able to gain a 40 percent premium for "quality" bananas in northern Europe! However, data indicated that despite Rainforest efforts, the savings were not passed back to producers (Chambron 2005, 103). SAN continues to have little dialogue with unions.

ISO 14001

In the mid-1990s, the International Organization for Standardization (ISO) developed a voluntary environmental-management system, ISO 14001. Third-party verification would determine whether organizations were in compliance with the system through their own auditing, performance, labeling, and life-cycle targets. Critics accused ISO 14001 of being a northern-driven, corporate design requiring few specific outcomes (see, e.g., Krut and Gleckman 1998), but some rated it as potentially "one of the most significant international initiatives for sustainable development" (IISD 1996). ISO 14001 subsequently evolved "objective criteria" for verifying claims (see Praxiom Research Group 2007). Their strict application in Ecuador caused certifiers to approve only 340 out of 13,500 farms! However skilled analysts still questioned ISO 14001's ambiguities.[1]

SA8000

Social Accountability International (SAI), a 1997 spinoff of the Council on Economic Priorities, adapted the ISO management system to aid firms control product quality by adding minimum regulations for respecting human rights—one reason its program director, Judy Gearhart, says she took the job. SAI creatively paralleled ISO accounting standards with an "impressive range for social policy, contracts, communication policy; working conditions for hired labor" (Chambron 2005, 99). SAI licensed firms like BVQT and SGS to conduct its audits. "Because earlier codes didn't mention freedom of association, SAI took a multi-stakeholder approach that included unions and NGOs . . . [,] instructing auditors to keep the standards tight in regard to eight ILO core principles: the four basics, freedom of association and bargaining, opposition to child labor, forced labor and discrimination; and four additional on health and safety, wages/hours, discipline and management systems." Gearhart believes that SAI offers handles in direct language that, if posted, workers can exploit beyond health and safety committees. SA8000 also provides incentives for managers to show what they have achieved to promote dialogue in the workplace. "We hope to make global competition about social compliance, not just about product, price and time!" While SAI is applied at farm and factory levels, it needs a supermarket result "to do with corporate culture. Various companies—Ahold and Costco are receptive. Can we sell this so that others don't get off the hook?" (Gearhart, 2005).

Chiquita chose SAI in 2000 to oversee the social and labor aspects of certification after honestly admitting that 45 percent of its banana farms were not in compliance (Chiquita Brands 2001; Taylor and Scharlin 2004, 187). Dole also used SAI, which approved its bananas in the Philippines and Costa Rica. Overall, SAI's cost-benefit analysis showed notable improvements in quality production, productivity increases, better worker morale, and higher revenue. Workers had fewer accidents, safer conditions, less discrimination, and more money for basic needs (Chambron 2005, 102). However, an FAO study determined that implementing SAI can increase costs dramatically, depending on how living wage or housing conditions are specified (FAO 2004).

Ethical Trade

Ethical Trade Initiative (ETI), a third TNC-utilized program, was created in 1998 by the United Kingdom's Blair administration with representation from unions, NGOs, and business to ensure that suppliers comply with environmental and social requirements, including a Base Code of Labor Conduct. Early on, the ETI Working Group on Freedom of Association and Auditing stressed the importance of reasonable hours, the voluntary nature of overtime, and adequate wages. Chiquita and Fyffes became ETI members in 2002. However, ETI does monitor compliance. In 2008, ETI and SAI signed an understanding by which SAI would utilize ETI findings.

IFOAM

The International Federation of Organic Agricultural Movements grew out of a 1972 effort among French farmers to promote healthy, pesticide-free produce. By 2007, it involved 750 organizations from 108 countries (http://www.ifoam.org). Its definition of what constitutes "organic" has been continually contested and revised but its committees on standards and accreditation monitor certifications. After 2000, IFOAM struggled to include social justice as an "integral part of organic agriculture and processing." (Shreck 2004, 11ff). It collaborated with FAO, FLO, SAI, and SAN in a pilot audit of a large banana plantation in Costa Rica. However, by 2008, IFOAM had yet to ratify its own proposed "Chapter 8," which required individual producers to state their human rights policy, prohibit forced labor and discrimination, offer children educational opportunities, and guarantee the right to organize and bargain.

Supermarket Codes

As retail giants consolidated control over commodity marketing in recent decades, they developed their own codes for certifying the products they handled. To counter bioterror threats, officials set baseline controls that would enhance food safety by quickly identifying product origin and scope. This also enabled companies to narrow any health threat and avoid having to cancel all exports from a particular region or country. U.S. antiterrorism measures likewise demanded an identifier that could trace each box of bananas to a specific plantation.[2] In 2003 twenty-two EU firms set up the Euro-Retailer Produce Good Agricultural Practices, or Eurepgap code. Besides traceability, it emphasized sanitary and phytosanitary norms, and some pesticide and health security rules, but it did not speak about labor rights (Chambron 2005, 97). By 2007 Eurepgap had certified over a hundred accreditation bodies to test fourteen food safety areas in more than seventy countries, including worker health (http://www.eurep.org). Some retailers introduced a laser process for labeling each piece of fruit. Others joined the Business Social Compliance Initiative, loosely based on SAI, or the much weaker Global Awareness project sponsored by Wal-Mart. Several promoted their own codes.

At mid-decade, Whole Foods, the largest natural foods retailer in the United States, Canada, and the United Kingdom ($5.6 billion in sales and 183 stores, as of 2006), formed a certification relationship and "Whole Trade guarantee" with Earth University in Costa Rica. Founded in 1990, the university, and its forty professors and four hundred students, viewed itself as "the leading tropical agricultural school for the developing world" (http://www.earth-brand.org). Earth University certifies pineapples and bananas on its own farms, where it uses organic postharvest fungicides, chemical-free bags, and no herbicides, and it recycles relevant wastes. To meet social criteria, it claims to pay full-time employees wages and benefits "above the industry average." Whole Foods also features some products endorsed by the Rainforest Alliance. While Whole Foods prides itself on bringing fresh and healthy produce to northern consumers, it is notably antiunion.[3] Its Earth certification does not reference specific SAN or ILO benchmarks, leaving undefined the conditions of its application of aerial pesticides, its actual wage payments, and the rights of non-full-time employees. NGO fair trader Jonathan Rosenthal admires Earth University's sustainable-development efforts in Costa Rica. Yet he believes Whole Foods has solicited Earth certification partly as a defensive measure: "Whole Foods, like Starbucks, worries that a multistakeholder approach could jeopardize its control. To avoid this, they say, 'We set up an umbrella program for checking our products that may in various places or times utilize four or five different methods of certification.' If one becomes too hot, they can go to the other; and over time, they may gradually incorporate various principles from all of them" (interview with Rosenthal 2007).

Code Combinations

Although they generally distinguish their products from those approved by Fair Trade,[4] TNCs use an *assortment* of certifications to advertise their bananas as "high quality": one third-party program to verify ecologically responsible productive methods, another to check socially responsible approaches, a third to please retailers. In 2005 Chiquita, which already enjoyed Rainforest endorsement, claimed 100 percent SA8000 certification for its own farms, and Eurepgap approval for its operations in Colombia, Costa Rica, and Panama. "We always have to meet more requirements," explained Chiquita's Vice President Enrique Vázquez. "We can't say no" (interview with Vázquez 2004). Dole joined SAI's board in 1999, combining the SA8000 code with its use of ISO 14001 and Eurepgap. By 2001, Dole had conducted at least twenty-five audits in five nations, investing several million dollars in warehouses, maintenance facilities, box plants, packing plants, and reforestation (Taylor and Scharlin 2004, 181). Dole "needs certification to demonstrate to our customers that we meet international standards" asserted Vice President Sylvain Cuperlier, who hoped "to have all our independent producers certified by the end of 2005" (Cuperlier 2005). Several studies evaluate firm performance on the various criteria (Chambron 2005, 98). Chiquita and Dole sought to link the ETI base code with SA8000 standards, earning high ratings. Del Monte gained full Eurepgap approval in 2003 but otherwise employed only ISO14001. Independent appraisers rated the company's transparency and adherence to social criteria unacceptable (Ethibel 2005).

The Eurepgap standard has caused dissension, not over content but over control. TNCs desirous of retail market access have felt forced to become certified by Eurepgap. Chris Wille, of Rainforest/SAN, projected that the Eurepgap obligation would have a bigger impact on banana sales than trade quota losses. Wille feared that "these rules will create a meaningless bureaucracy" with no guarantee of higher prices for producers. Since the Eurepgap code did not cover buying practices, Wille feared supermarket sourcing could be unethical. Rainforest hoped "to keep NGO certifiers [like themselves] in the lead and not relinquish control to retailers" (interview with Wille 2004).

Yet the frenzy over codes appears to have enhanced consumer awareness about the environmental impact and strenuous work conditions of banana production. While full documentation is in order, considerable media interest in codes has correlated with increased sales of certified fruit. Code components may thus contribute to a "responsive" consumer identity, shown by purchasing "just bananas" along with "sweat-free" clothing and "fair-trade" coffee. Nevertheless, farmers and workers express caveats about how this identity might by altered by other promotion programs.

National Banana Promotion Programs

Local producer associations have joined the certification fray to promote their export quality. Costa Rica, Colombia, Ecuador, and others have each implemented promotional efforts abroad. Costa Rica's Banana Export Association (CORBANA) exemplifies how national and TNC firms collaborate to solve common problems such as the persistent black sigatoka leaf disease but also to market (i.e., promote) the country's competitive advantage. The latter raises questions about CORBANA's disregard of labor issues.

The CORBANA Promotion

In 2000 CORBANA inaugurated a campaign to promote the image of Costa Rican bananas in the United Kingdom, Germany, and Belgium "as a consequence of the loss of quotas and the many attacks that the Costa Rican banana industry had faced from NGOS in Europe" (CORBANA 2004).[5] Adopting the slogan *"el mejor del mundo"* (the best in the world), the Costa Rican banana industry focused on

- the most important supermarket chains
- NGOs devoted to environment and social conditions
- public media (press)
- consumers (indirectly via opinion makers).

CORBANA's message was: "Costa Rican bananas are different. They are produced under friendly social and environmental conditions." The organization sought to pitch the idea that, "while it was important how Costa Rican bananas were perceived . . . it was most important how they were produced." Over the next three years, CORBANA representatives visited fruit and vegetable fairs in Europe, lunched with businessmen, wooed press attention with video displays, and solicited NGOs, supermarkets, and politicians. In Belgium, they sponsored a bicycle event, the Vélo de Ravel, a five-week-long run where CORBANA furnished T-shirts and free bananas. In the United Kingdom, they "sought out representatives of ACP countries and established a potential alliance in 2006"! They also convened leaders from Fair Trade and Friends of the Earth, on whom they claimed "great impact. Both groups acknowledged that they had never contacted or known about the industry due to the denunciations made by unions" (CORBANA 2004, 19). This assessment might explain why CORBANA has been reluctant to promote union rights.

But CORBANA believed it had opened the door to these groups: "Importers and supermarket representatives spontaneously expressed that Costa Rican bananas are the best in the world. . . . Attacks by NGOs have been considerably reduced. Industry filled the empty space proactively with open doors to key NGOs in Europe, especially in England and Germany" (CORBANA 2004, 22). CORBANA Director Martin Zúñega was ecstatic: "The Europeans especially are insisting on international transparency, information, and certification as strong elements of consumer satisfaction. We stress that Costa Rican bananas use less toxic chemicals, and we emphasize our environmental and worker protections. We expect that in the next eighteen months, 100 percent of our plantations will be certified." As a result, Costa Rica "actually increased our European market share in 2003 to 28 percent from 25 percent in 2002. We think our campaign and environmental approach is partially responsible for our 3.4 percent increase. Although we dropped in land area planted from 50,300 hectares in the early 1990s to 42,000 in 2003, we are gaining in productivity. Whatever trade scheme prevails, we can still make the argument on quality and the manner in which we produce bananas" (interview with Zúñega 2004).

Other countries have engaged in similar national promotion campaigns. With government help, Colombia advocates "Banatura," stressing respectful social and environmental conditions on fincas that use its personal insignia. Ecuador's Corpei export association actively touts bananas as the "Producto de Ecuador." However, if national producers disregard ILO principles while their national campaign promotes their fruit's production as attentive to social criteria, the campaign's efforts will contravene worker protections.

In fact, despite the growth of third-party certification as a marketing strategy by TNCs, national firms, and quasi-government entities, the actual impact that certification has on workers is largely undocumented. Case studies offer very limited evidence of positive effects. Chiquita's SAN project in Costa Rica conveyed benefits with only a 1.5 percent increase in cost (FAO 2004). However, field verification on yields and profitability remained difficult (see Chambron 2005). A Swedish comparative study pointedly questioned whether bananas could be grown sustainably via a monocrop system: despite phytosanitary rules and inspections, black sigatoka and crown rot still required a significant amount of nematicides, fungicides, and insecticides (Mattsson 2004). The national producer associations published certain benchmarks for meeting labor and social standards but offered no data to verify their having been reached. COR-BANA's program to contravene union "denunciations" to European NGOs illustrated why labor remained skeptical about any certification scheme—whether it emanated from TNCs, their counterpart certifiers, supermarkets, or national producer associations.

Labor Perspectives on Certification

Workers do view the industry's quality initiatives as an opportunity since they could help brake a race-to-the-bottom banana market. Nevertheless, they caution consumers that codes deceive in claiming better social conditions, especially the employees' right to free association. Despite code references to ILO Conventions 87 and 98, companies routinely interfere with workers' attempts to form unions and negotiate contracts. At the very time companies emphasized code promulgation, they also deployed severe threats and bribery to challenge unions. Occasionally they appeared to reasonably negotiate. Where unions existed, some plantation officials agreed to improvements, although others demanded givebacks. But "in plantations where there was no trade union at the time of certification, certification has not led to freedom of association" (Chambron 2005, 105). Instead, workers experienced stringent TNC resistance: Dole supported antiunion Solidarista associations and threatened to leave Central America if European trade rules become too costly. Chiquita flirted with breaking its union Global Framework Accord. Del Monte broke its union-access agreement in Costa Rica and cut wages by 40 percent. Fyffes shabbily treated its workers in Belize to lower wages and benefits (Chambron 2005, 104). If unions demonstrated strength, codes did not prevent companies from moving to nonunionized locations where work was cheap. Dole sold its Colombian, Nicaraguan, and Venezuelan operations, all of which were unionized, choosing to source nearly half of its bananas from nonunion Ecuador and Guatemala's Pacific coast. Chiquita shifted production to Angola and Guatemala's southern coast, where it acknowledged independent suppliers often do not respect freedom of association (McLaughlin 2006). Del Monte expanded in nonunionized areas of Cameroon and Brazil.

This evasion angered workers trained to improve quality and productivity. While TNCs promoted Latin bananas as "justly produced,"

TABLE 11.2. Union Recommendations for Effective Codes

Code Aspects	Worker Impact
Truly reflects ILO rules	Promotes free association, bargaining
Complements, reinforces local law	Strengthens inspections and courts
Utilizes worker-sensitive monitors	Properly pictures labor conditions
Emphasizes ongoing worker evaluation	Accurately portrays consistent realities

they cut wages, fired workers, and thwarted unions. Leaders in Central America and elsewhere quickly perceived class dynamics to undergird codes of conduct, and workers quickly decided that industry exploited codes to bolster its international image, not to address employee needs. As the secretary general of SITRABI's Confederación Unidad Sindical de Guatemala (CUSG), put it, "Companies supposedly have rules protecting workers, but I don't know of any company using them to help improve conditions. I believe they are really utilized to justify an image at the international level" (Mancilla 2001). Still, union activist networks also grasped the potential opportunity of international-marketing vulnerability, or what Keck and Sikkink (1998) called the "boomerang effect" of global leverage on local firms.

Naomi Klein has demonstrated how noted corporations popularize a lifestyle to accompany their identifiable brands (2000); workers viscerally understand the TNC strategy of advertising products in a global market to encourage middle-class consumption. They chuckle when companies actually cite work conditions as a marketing ploy (e.g., Starbucks displaying Guatemalan women in multicolored indigenous *traje* picking coffee). They appreciate northern student and consumer protests over corporate exploitation and the irony of quality products grown by workers who barely get enough to eat; and they understand how codes could be a benefit in addressing these disparities. As the Guatemalan union UNSITRAGUA acknowledged, "If a label is well known, the certification process may provide us with some pressure" (interview with Barrientos 2001). Costa Rican and Honduran unions also cited code violations to generate international scrutiny.

However, unions are clear that codes can help only under careful conditions, summarized in table 11.2.

Codes Must Reflect ILO Rules

Many codes used for banana certification (table 11.1) contain references to the core ILO standards briefly outlined in chapter 8. Incorporating these references did not occur automatically. It emanated from worker, student, and consumer pressure that demanded the explicit designation of rights in predecessor codes. Most basic issues were originally debated under the ILO's tripartite business, government, and labor auspices. Labor history demonstrates how workers tenaciously petitioned to gain the inclusion of key protections (see Haufler 2001; Hopkins 1998; Murray 1998; and Thamotheram 2000; see Vogel 1978; Williams 2000; and Willmott 2001 on corporate codes).

Once labor protections are embodied in the codes, the challenge rests with their interpretation. Unions fear that codes make private monitors, not the international agency, the new arbiter, "as if they conceived the ILO rules" (COLSIBA 2007). Most contention occurs over freedom of association, which I will discuss in chapter 12, but other disputes arise as well. In 2006, Costa Rican unions reported that SAI and Rainforest monitors stepped up weekly visits as TNC producers, hoping to counteract unions, increasingly subcontracted production to independent producers with poor, often illegal work conditions. One Rainforest report allegedly informed Chiquita that workers at Caribana's Costa Rican plantations "were happy and only working eight hours." COLSIBA charged that SAI and Rainforest compliance reports glossed over collective-bargaining and harassment violations, while SAI indicated it did not have sufficient budget to investigate charges submitted by unions. In turn, unions demanded a greater voice in determining ILO interpretation and enforcement.

Codes Must Give Primacy to Local Laws

Labor officials worry that certification codes are being substituted for legislation and want this reversed. Central American countries have detailed rules for recognizing unions and preventing discrimination and child labor. National rules control minimum wages and procedures for health insurance, pensions, and women's rights. Laws protect workers who file for union status or who submit grievances to the labor ministry regarding a particular employer. Yet time and again, workers who file complaints or attempt to achieve union status are quickly identified and fired. Agricultural workers face enormous obstacles arranging for time and travel to attend court appeals. Even when an occasional courageous judge orders union recognition or reinstallation of fired workers, companies do not comply. The private sector roundly criticizes the incompetent labor ministry and courts but eschews taxes and fines that would provide the ministry with funds to carry out its responsibilities. It prefers to bribe poorly funded labor inspectors and court officials to overlook or withdraw regulations that demand fair treatment. From the workers' perspective, the problem is not a lack of codes or regulations; the problem is that the companies do not obey the laws already in force.

Workers also find current codes even less effective than national labor legislation, which is why law-violating companies propose private versions of the regulations they flout. Firms can then hire their own certifying agents to validate compliance. To workers, this is even worse than paying off labor inspectors or judges for a beneficial ruling, since it further undercuts the state's ability to implement its own laws. For UNSITRAGUA, "companies certifying other firms is nothing more than a *privatization* of the labor code. The only way to do certification is to respect labor laws, and follow the mandated rules" (emphasis in the original) (interview with Barrientos 2001; also Reich 2007). Leaders urge that national legislation be made more specific to avoid being replaced by certification schemes that do not have the legal force of the state (Harari 2005b).

Nevertheless, labor networks familiar with trade pressures recognize that codes can be used to enhance local labor compliance by pressuring name-brand firms to respect the process, probing labor ministries to do their job (see Frundt 1998). True, the law should be sufficient; but when armed with company codes that index local laws, unions may find it easier to gain a more objective interpretation and implementation of the law.

Codes Must Be Audited by Worker-Sensitive Monitors

Auditors accredited to verify compliance with ISO 14001, SA8000, Eurepgap guidelines, and so on, are usually one of three types: transnational for-profit firms like Ernst and Young; transnational nonprofit firms like Verité, SGS, or Bureau Veritas Certification (formerly BVQI); and local nonprofit groups of human rights and academic activists.

Local auditor teams bring advantages but also baggage. Some international unions, such as the International Textile, Garment and Leather Workers Federation (ITGLWF), which until 2006 served on the SAI Advisory Board, are wary. "Are NGOs better than professional auditors without training in production systems, industrial relations, health, wages and accounting?" asked Neil Kearney, ITGLWF secretary general (interview with author, 2002).

No matter the type, since the brand firms and/or the local producers hire the monitors, workers believe monitors protect subcontractors.[6] Yet objectivity may also be affected by other factors. For-profit monitors are routinely drawn from business schools and accounting firms with almost no labor experience. Local NGO monitors from religious, human rights, university, and small-business backgrounds are virtually always middle class. They may know about the country's political situation and share a commitment to democratic principles. However, they often have little direct experience with unions and their particular difficulties in a hostile environment. Nevertheless, as Coats (1998), Compa (2001), Esbenshade (2004), and Seidman (2007) have demonstrated, monitors may see themselves in an advantageous position to determine what is best for the workers. Auditors freely consult with management on worker issues without union presence, and they openly criticize the way certain unions handle their proposals. Most important, they often do not understand the primacy of free unionization over other sorts of rights. Instead, they privilege touted company policies that "maintain competition and jobs."

Monitors are also generally untrained in conducting confidential interviews in ways that bypass class distinctions (Rose 2000), engender worker confidence and assure anonymous complaints. Sylvain Cuperlier, Dole representative on the SAI advisory board, agrees that appraisals "are more difficult . . . because of greater subjectivity." He advocates "worker interviews to determine whether the workers have representation," yet he understands they offer "no guarantee to measure the truth, and it is not acquired by a team overnight." This requires "someone who has experience with unions" (Cuperlier 2005). As a code, SAI has made a notable effort to properly audit workers, despite its mistakes in Costa

Rica. SAI selects a diverse team that reflects expertise on health, safety, and worker representation to "reach enough facilities to show progress." Program director Gearhart is clear: "If we can't do it at a certain level, then it is just an experiment" (Gearhart 2005).

Codes Must Assure Ongoing Worker Input

In addition to audit interviews, certification must include mechanisms for employees to offer frequent and consistent independent assessments. Speaking of the Rainforest Alliance and SA 8000 codes applied at Chiquita banana fincas, the head of the international relations at UNSITRAGUA found that "management really spruced things up the week before the certifiers arrived—cleaned up the garbage, stopped aerial fumigation, took chlorine out of the rinsing baths—and all seemed fine. But once the certifiers left, they went right back to doing it the way they did it before. They are just hiring another private firm to 'paint their house'" (interview with Barrientes 2001; see also Bendell 2000).

To avoid this, codes must be widely promulgated and understood. In fact, few workers know about codes, so they become a "missed opportunity" that labor leaders believe is intentional. David Morales, secretary general of the Guatemalan Food Worker Federation (Federación Sindical de Trabajadores de la Alimentación, Agroindustria y Similares, FESTRAS) points out that "some workers are familiar with codes. Certainly, the brand is a key factor in our favor. But they can be just a piece of paper in that we don't know how to use them" (interview with Morales 2001).

In the judgment of Canada's well-respected Maquila Solidarity Network, SAI has developed the most elaborate code for assuring parallel worker representation and anonymous complaint procedures (MSN 2001). SAI's Judy Gearhart assures that "management has the responsibility for communicating to workers what is in the code" and how employees can utilize it. Tests and worker training can be helpful, "but how can there be tests and dialogue if workers can't read the codes," which are often only in English (Gearhart 2005). Some codes are available on Web sites, but they must be explained to unions and made accessible to plantation workers. It takes a network with special resources for unions to utilize such codes effectively.

To aid worker instruction in 2001, the ITGLWF inaugurated a joint program with SAI to train southern workers in filing complaints and reporting unjustified certifications (MSN 2001). Five years later, ITGLWF and MSN judged it had not succeeded (it also terminated SAI board participation). When one Norwegian union tried to follow up on a 2005 SA8000 certification complaint in Costa Rica, it received no reply from SAI (Chambron 2005, 105). SAI's third-party audits, often done by for-profit companies, are not made public.[7] The nonprofit certifier has created a special "Signatory Program" through which companies offer more public information and assist subcontractors in reaching compliance (see also Leipziger 2001; MSN 2002), but the process has yet to be perfected. In chapter 12 I further consider why the only effective monitoring includes open, daily reports by a democratically elected union.

Summary

Third-party banana certifications have proliferated, been critiqued, and generated alternatives. Nevertheless, codes have sparked interest among consumers. In the late 1990s their application in the apparel sector became a cause célèbre among college youth that spread to a wider public. As ethical worries expanded to agricultural products, TNCs devoted more resources to promoting their image of quality on the basis of social dimensions and environmental commitment. Their spotty achievements created an impetus for network action and consumer identity with fairer trade.

The certification process thereby presented a structural opportunity for workers, small farmers, and consumers to forge common understandings and networks around just what ethical purchasing entails and how they must be part of the process. This opportunity remains tenuous: free-market policies continually stimulate firms to squeeze third-world labor, resulting in the double-bladed message that code certification can muddle.

Fair Trade's attention to bananas represents a dramatic effort at clarity to prevent this squeeze. Going beyond most certification programs, FLO labeling addresses the major decline in prices banana producers receive. It invites consumers to advance the opportunity for just livelihoods, sustainable practices, and free labor participation. To this possibility, we now turn.

CHAPTER 12

Fair Trade and Freedom of Association

IN 2005 COLSIBA and TransFairUSA jointly announced their collaboration for "the development and growth of the market of Fair Trade union bananas" (TransFair/COLSIBA 2005). To supporters, the announcement trumpeted the alliance of three very important banana constituencies: trade-conscious consumers, banana workers, and small farmers. It also marked a major step in the decade-long effort to resolve a key issue faced by all certification schemes: how to implement the essential principle of freedom of association (FOA) as an indicator of "quality bananas."

As a certification program, FLO offered unions a special opportunity because it includes two mechanisms that addressed worker rights, outlined in chapter 3. First, it promised a rigorous audit to ascertain genuine freedom of association. Second, it demanded a social premium that must be distributed democratically via a collective process. Labor leaders recognize that these mechanisms could encourage the formation of unions within the intent of ILO Conventions 87 and 98. However, the leaders also caution about their potential misuse. Following union recommendations I set out in chapter 11, in this chapter I pursue why freedom of association is viewed as the most important of all union principles. I then discuss ways in which this right was encoded by the ILO and other bodies. I further elucidate why unions often object to the way inspectors and certifiers apply the principle. Concretely, I point to some of the issues FLO, TransFair, and COLSIBA have faced in their efforts to uphold freedom of association in specific FT-certified Latin cooperatives.

Freedom of Association

The right of working people to form independent associations has been recognized since the time of medieval guilds. While governments would invoke restraint-of-trade rules to thwart bakers, tailors, and others from combining to set prices, the workers repulsed these efforts.[1] In 1884, they gained a legal basis for the right of association in the Waldeck Rosseau Law of France. It guaranteed workers the ability to associate independently and to choose spokespersons who would represent their interests in a collective contract with business owners. While the ILO championed this right ever since its founding in 1919, it officially ratified free association and bargaining in 1948–49, contemporaneously with the U.N. Universal Declaration of Human Rights. The ILO's Conventions 87 and 98 codified the right to organize independent unions and to bargain collectively as the most basic and far-reaching among its principles (see Dunning 1998; Gernigon, Odero, and Guido 2000; Swepston 1998; Potobsky 1998; and Valticos 1998).

Religious organizations and political entities have likewise advocated the rights of association and bargaining for independent trade unions. Fearing socialism, the Roman Catholic Church proclaimed such rights via its papal encyclicals on social justice such as Pope Leo XIII's *Rerum Novarum*, in 1891, and Pope Pius XI's *Quadragesimo Anno*, in 1931. The U.S. government incorporated such principles into law when it legalized trade unions with the Wagner Act, in 1935, although it exempted agricultural workers. In the aftermath of World War II, U.N. delegates believed the legal formation of independent worker associations would help inoculate against communist movements, and in 1949, the U.N. chose the ILO as its enforcing agency. Fundamental to all these expressions were the employee voices that kept pressuring for workplace rights. At the insistence of the World Federation of Trade Unions, the right of association offered legitimacy to independent worker organizations within a capitalistic economic milieu (French 1994).

Most nations of the Western Hemisphere subsequently ratified the U.N. Universal Declaration of Human Rights and two adjacent covenants on economic and social rights that embody the universal right of association. Nations that retained ILO membership also implicitly agreed to abide by ILO core conventions even when not explicitly ratified.[2] Before World War II, most banana-producing countries hesitated about adopting the principles embodied in Conventions 89 and 98, which linked free association to collective bargaining and the right to strike. In 1943, Costa Rica became the first Central American nation to ratify a labor code that embodied these provisions. Subsequently, the region's governments passed roughly similar labor legislation. Variations concerned the number of workers required to form a union (from twelve to forty), the voting percentage of the workforce needed to present contract demands to management (from 25 to 51 percent), and the percentage necessary to approve a legal strike (from 50 to 67 percent). Underlying these differences was a growing understanding that the right to associate without the power to negotiate was meaningless. Only when the workers could democratically choose representatives who presented their issues to management face-to-face, and be guaranteed a bona fide response that they could accept or reject within a reasonable time period, would they have effective independence. Their power to negotiate occurred when they had the ability to withhold services.

The designers of ILO principles understood that establishing detailed international standards for working conditions is difficult: wages, for example, can vary substantially by region. They likewise foresaw problems in legislating common procedures for monitoring safety conditions. Cultural differences would affect how basic expectations could be best understood or applied within a globally competitive economic environment. Given a general framework of rights, the designers therefore recognized the organizations closest at hand as being in the most advantageous position to assess appropriate wages and working conditions for their own situations. Local unions could best negotiate these outcomes with employers, at times with government input. For this reason the right of association at the local level retains preeminence among other socioeconomic rights.

Thus, freedom of association's essential value is that it places primary responsibility for solutions at the level where it can be most effective — with the people who are most familiar with the direct impact of work conditions. It effectively positions the principles of subsidiarity, participation, and democratic decision making of those directly affected alongside the comparative advantage of producers. By guaranteeing an equitable assessment through give-and-take negotiations, the principle maintains flexibility for both workers and employers in how they apply other core ILO conventions opposing slave and child labor and discrimination, and favoring health and safety and adequate wages. Workers in free associations also make these determinations more easily because they pay their own way and are financially independent of management's influence.

Free Association in Practice

Despite a history of more than sixty years of universally validating the right of association, governments and companies have often stymied its implementation. They have engaged in a variety of subterfuges, such as delaying recognition, subdividing the workforce, encouraging a plethora of worker organizations to dilute required numbers, recognizing unions but refusing to bargain, or readying replacement workers to prevent strikes. On the one hand, in the latter twentieth century, as military governments increased power in banana-producing areas such as Honduras and Guatemala, they brutally attacked unions (see Burbach and Flynn 1980; Jonas 1991; Peckenham and Street 1985; Armstrong and Shenk 1982; and Striffler and Moberg 2003). Yet until the early 1990s, the U.S. State Department and U.S. Trade Representative (USTR) refused to link violations of the right of association with violations of the right to life.[3] On the other hand, certain authoritarian leaders, as in Panama, Colombia, and Somoza's Nicaragua, built alliances with trade unions to maintain state control, although labor's power was never equitable. While unions might have won some wage improvements, they did not gain an autonomous voice (Wirada and Kline 1985; Collier and Collier 1991). We also saw corporate efforts to replace unions with employer-controlled Solidarista associations that still persist in new guises such as management-labor "joint bodies" that beguile well-intentioned NGOs. In the twenty-first century, the most common antiunion strategy has been to simply fire union members and leaders, as Chiquita, Dole, and Del Monte still allow in Central America.[4]

Certifications and Freedom of Association

While nearly all banana-certification programs index ILO Conventions 87 and 98 recognizing association and bargaining,[5] unions state that third-party auditors are often untrained in assessing union freedom. As illustrated in the last chapter, auditors fall short in applying the principles. They have only limited, periodic discussion with workers. Some are content to simply ask managers, "Are employees here free to associate?" Others perfunctorily determine freedom's presence by whether or not an employee association exists and holds occasional meetings. Monitors do not understand that workers must have a protected way of deciding their own representatives and of expressing their complaints to management. While free association does not require a union, since workers can choose not to have one, auditors do not comprehend that workers often select the "no union" choice under duress. Certifiers are not properly instructed in how to put workers at ease so employees can talk without repercussion. SAI Program Director Judy Gearhart acknowledged that "workers were not left alone to speak, and when asked about it off premises, they give very different accounts," motivating SAI to establish its parallel complaint mechanisms (Gearhart 2005).

Auditors must learn how to use multiple indicators to detect free association. Deciding if such an employee association is truly an independent voice for workers involves probing into legal and social questions. In certain countries, like Mexico, companies clandestinely register union entities, and use "hidden" or "protection" contracts to prevent genuine union organizing (Alexander 1999; Bacon 2004). Elsewhere, auditors checking for free association have been documented accepting Solidarista associations as unions. Not only are they not unions, they have the elimination of unions as their objective (Flores 1993; Vado 1998). "The ILO has made very clear that Solidarista associations should not have a role representing workers. So when certifiers have decided that plantations [with Solidarista associations] are OK, from the IUF view they may be . . . undermining freedom of association" (Longley 2005). SA8000 has "clarified that Solidarista presence in Costa Rica is a violation of freedom of association. It is not a legitimate choice" (Gearhart 2005). Companies have responded by seeking an ILO exemption.

Unions invoke the principle of free association to argue why *they* should play a role in the certification process. "How does one decide if a free and independent union exists?" queries the IUF's Sue Longley. Auditors can obtain only a "single snapshot" of what is happening. "It may take us several years to make this determination" (Longley 2005). Since the "snapshot" does not present the whole picture, someone has to fill out the details. SAI's Judy Gearhart does not dispute the workers' viewpoint that the best monitors would be those present each day who can collectively pool their information to present an accurate and complete picture of what goes on. She acknowledges the need "for more union input into the auditing process" (Gearhart 2005).

Honduran banana unions illustrated this point in early 2005. They persuaded Chiquita to use plastic bags without the Durisban (chlorpyriphos) insecticide coating. However, an independent plantation supplying Chiquita refused to go along. In April 2005 forty-eight members of the affected SITRAAMERIBI/SITRATERCO union went on strike to demand the pesticide-free bags. The company attempted to declare the strike illegal and fired the forty-eight workers. Because of Chiquita's policy, the workers were quickly reinstated, and the national producer agreed to adopt untreated bags (COSIBAH 2005). Yet had Chiquita responded to immediate worker complaints it could have prevented things from getting out of hand. "Since they are on the ground everyday, trade unions could play a key role in inspections, auditing and monitoring," concluded Longley (2005).

It is true, of course, that union officials, like any organization's officers, can be swayed by political considerations and at times corrupted. They can make special arrangements with managers to present a flattering picture; or they can illegitimately threaten to publicize an unfairly negative report. Likewise, "there is also corruption among small farmers," comments Jonathan Rosenthal, Fair Trade banana consultant. "We have to think outside the box to create real empowerment" (interview 2007). Monitors can establish safeguards to promote empowerment, but the best way to guarantee it is the workers' collective knowledge: their honest presentation of conditions gives them the best assurance of fair compensation and future job security.

Fair Trade and Freedom of Association

As I emphasized earlier, the Fair Trade movement's original objective was to improve living standards, first for small producers, later for plantation workers, and eventually for everyone in the supply chain. In 2003 FLO-CERT became a legally independent body responsible for conducting audits on all FLO-related producers, fielding eighty-five local inspectors by 2007.[6] The FLO/FLO-CERT approach has won recognition for preserving biodiversity, preventing soil depletion, and handling waste. Academic comparisons have given FLO's cost-accounting measures high ratings for assuring workers a livable wage (Mattsson 2004). Gaining 1 percent of world banana sales in 2006, FLO producers earned an extra $30 million they otherwise would not have (FLO 2007). Likewise, an FAO study demonstrated that besides FLO's price premium, its benefits included better organization, bargaining position, creditworthiness, capacity building, guaranteed markets, access to international markets, and learning-by-doing export programs. All this occurred despite the fact that in some cases FLO supply was higher than demand, so participation did not always directly improve producer income (Chambron 2005, 102; Shreck 2002, 2005).

Unions quickly grasped the benefits of Fair Trade and joined its early promoters. Yet they also feared that organizations that founded FLO and those importing FLO bananas suffered the same lacunae regarding the principle of free association and ILO Conventions 89 and 98 as other social certifiers. European FLO organizations held divergent positions on trade policy and had no unanimity about "worker empowerment" or how to evaluate claimed violations of freedom of association. Irish Fair Trade leaders were outspoken on behalf of labor rights (Lamb 2008, 51, 111). Yet in certain instances, FLO certified fincas hostile to unions where questionable labor practices endured. "Some cooperatives are not genuine cooperatives" (Longley 2005), and Fair Trade's two-tiered basic/advanced requirements appear to have delayed the application of criteria in a way that enabled organic growers and smallholders to hire landless workers for very low wages. By not giving sufficient attention to labor rights, while encouraging consumer purchases of such Fair Trade produce, FLO appeared to the unions to be undercutting union-produced bananas.[7]

Fair Trade Cooperatives and Hired Labor

We found some basis for this fear among Latin Fair Trade banana producers. Table 12.1 reveals wide variation among countries in the number of their associations, members, and permanent and seasonal workers; nevertheless, (excluding Ghana and the Windward Islands) two-thirds of those involved in production are *hired* workers, accentuating the importance of the free-association question.

TABLE 12.1. Fair Trade Banana Production, Producers, and Employees by Country, 2006

Country	2005 Production (metric tons)	Associations	Permanent Workers	Seasonal Workers	Members
Colombia	4,620	15	1,095	363	145
Costa Rica	2,310	1	170	2	70
Dominican Republic	17,197	18	1,758	135	982
Ecuador	20,020	8	891	1,323	652
Ghana		1	572	17	
Peru	1,540	6	15	140	1,203
Windward Islands (St. Vincent)	5,134	1			628

Source: FLO 2007; production from Raynolds 2007.
As an FT source, Ecuador has subsequently lost ground to the Dominican Republic and Colombia.

Freedom of Association in the Caribbean and Central America

The WINFA/FLO collaboration in the Windward Islands is given high marks for freedom of association, despite those difficulties described in chapter 9. FLO was the only certification to directly address commodity price declines, enabling eighteen hundred small producers to survive (Chambron 2005, 106). At the same time, it encouraged and maintained complementary relations with unions (Longley 2005). In the Dominican Republic, however, the remarkable expansion of organic and Fair Trade production placed increased demands on farmers, who in turn exploited workers. Although on a temporary basis, 66.7 percent of Finca 6 members utilized hired labor.[8] According to German Zepeda, its exporter Savid contracted "half [of its] output to be picked by low-paid Haitians.... 80 percent of employees on three Savid farms are outside workers." In 2006, Savid finally met with the unions "under the auspices of TransFair" (interview with Zepeda 2006). On a Dominican visit in 2007, a Swedish Food union came across a certified FT banana plantation with three hundred workers.

In Costa Rica, the situation was mixed. Although not officially endorsed by FLO, Talamanca's Association of Small Producers (Asociación de Pequeños Productores de Talamanca, APTTA) on Costa Rica's Atlantic coast offered a unique example of freedom. APTTA represents 1,067 members from thirty indigenous Bribrí communities, 38 percent of whom are women (http://www.appta.org). It had a committed tradition of equality; however, fresh shipments of its smaller fruit have not reached a large market (McCracken 2002, 2004), totaling only three hundred boxes in 2006. Most of its output was used for manufacturing puree.[9]

However, CPASM de Trabajadores Bananeros del Sur R.L. (Coopetrabasur, or the Co-op), on Costa Rica's Pacific coast, presents a formidable FOA challenge. Founded about 1980 by Chiquita workers terminated when United Fruit moved to the Atlantic zone, the seventy-member cooperative began Fair Trade production in 1997 after struggling with earlier marketing difficulties. Members reduced chemicals, eliminated paraquat, recycled plastics, planted trees, and implemented a democratic structure. Wages increased, and the Co-op used its social premium to hire environmental specialists, make housing repairs, and "improve healthcare, schools . . . for banana producers and their families" (Arnett 2007). Founding member Arturo Jiménez Gómez extolled Fair Trade. "Before I was someone that took a box and loaded it onto a train . . . I was just a farmer, an intermediary. In this new system, I have become an international businessman." Jiménez hoped that his European friends would continue to buy Fair Trade bananas in a sustainable way so that he could become "free commercially, to have access to markets, to have the opportunity to dream of being free, to dream of being looked upon as a human being, not an object" (cited in Liddell 2000).

However, after demand increased in 2004, Coopetrabasur hired about ninety day laborers to meet shipments. The laborers received benefits but were not part of any union. When a COLSIBA-endorsed visitor from ASEPROLA sought to interview them and ascertain conditions in 2005, the Co-op refused entry. COLSIBA subsequently objected that FLO certification of Coopetrabasur bananas misrepresented rights compliance, and certain EUROBAN members, like CTM, ceased their purchase. In 2006, Coopetrabasur exacerbated the situation by purchasing the Algarba plantation and establishing a separate contractor, Cooperser-sur, to handle Coopetrabasur's labor requirements, infuriating COLSIBA.

Nevertheless, AgroFair and Oké maintained sourcing from the Co-op. In Ireland and the United Kingdom, Banana Link mobilized workers and the Fair Trade Foundation to make sure no plantation would be certified unless it had an active union.[10]

Fair Trade and the Andes

Likewise, FLO-approved Andean sources in Ecuador, Peru, and Colombia illustrate the benefits of Fair Trade for small-farm bananas but also the risks to freedom of association. In Ecuador, El Guabo, El Prieto, and several others grow Fair Trade bananas. Despite Ecuador's notoriously poor working conditions, OkéUSA believed its arrangement with El Guabo was enhancing a FT identity: For its smallholders, most farming less than thirteen acres (Lamb 2008, 16ff), the Co-op employed a "team of agronomists, technicians, and social workers to help farmers improve their business and ensure that Fair Trade standards are upheld." In addition to offering health coverage to its 220 producers and 66 field inspectors, "the cooperative provides regular health check-ups . . . sick pay and . . . maternity leave" for its 500 employees and 40 dockworkers. "In 2004 El Guabo used part of its Fair Trade premium to pay the school fees for its banana workers' children—444 in total. The cooperative also funds teaching materials, a bus service to bring children to and from school, lighting for classrooms, and a well with clean drinking water" (OkéUSA 2007). Additionally, it gave the children preventive health care. In 2007, under AgroFair auspices, Oké paid El Guabo $6.75 for an FOB forty-pound box of bananas and $8.50 for organic bananas plus a $1.00 premium.[11] Oké's Jonathan Rosenthal was in daily contact with and personally visited El Guabo several times a year: "If I observe stem-rot on a box of bananas, I want to be able to call up someone I know and ask, 'How can we remedy it?' We are a community of peers. We are joint owners. Whether farmers, farmer representatives, or marketing agents (like me), we eat together, go out to the countryside together" (interview 2007). El Guabo did hire part-time "quality checkers" whom it successfully encouraged to unionize as an independent organization. Fearing retaliation, however, the checkers could not mention their union to other companies with whom they were in contact (FTF 2007).

BanaFair distributes about three containers per week from UROCAL. According to Iván Ramón, its non-FLO supplier Co-op Cerro Azul was set up in 2002, with 138 growers. "Fair trade has increased the income of our members, decreased migration, provided employment opportunities to the community, and improved the self-esteem of small-scale banana producers" (TransFair 2005b). UROCAL then contacted COLSIBA.

Since 2000, the privately owned El Prieto has been shipping Fair Trade bananas with Dole as its marketing agent. By 2005, El Prieto's more than three hundred workers were producing 5,200 metric tons. About a third were hired labor (FLO 2005). Their complaints about low pay reached transnational activists who forced Dole to investigate (see chap. 13). By 2007, El Prieto had signed a memorandum of understanding but had not recognized any union.[12]

Unions contend that FLO-certified nonunion bananas from these and other Ecuadorian sources cost significantly less than union-produced bananas from Panama. One 750-acre Ecuadorian farm that supplied substantial Fair Trade shipments to Finland paid its 250 workers below-average wages. Women workers received even less than men. The wealthy owner pocketed the FLO premium to cover his insurance assessments from the Social Security institute.[13] "Yet workers fear to speak about unions," emphasizes German Zepeda of COLSIBA. "There is suspicion, not freedom of association for bargaining and striking." They insist that "*comercio justo no es justo*" (Fair Trade isn't fair) (2005).

In part for this reason, Ecuador, which had been the top FLO supplier, accounted for only 25 percent of its supply by 2007, with the Dominican Republic and Colombia assuming the lead. FLO approved the 157-member Colombian cooperative ASOPROBAN in 1998. By 2008, FLO had validated nineteen Colombian producer associations. However, association members represented only 10 percent of their workforces. Despite Colombia's trade-union tradition, certain FT co-ops prohibited their unionized members from serving in "joint bodies" and tapped the premium to conduct antiunion training and mischief with apparent FLO-CERT approval. However, FLO-CERT may have preempted some violations: in 2004, five non-FLO firms of the Magdalena region (Tropic S.A., Banapalm, Banex, Frutesa, and Agramayor) formed BANASAN to cultivate fifteen hundred hectares for an approximate production of at least 5,460,000 boxes per year, 30 percent of which was organic. BANASAN struggled to gain certification from EurepGap and RainForest. However, it was unsuccessful with Fair Trade. BANASAN's joint labor-management committees did not assure worker independence, and it purchased 25 percent of its supply from independent producers (http://banasan@banasan.com.co).

Peru's seven small FLO producers, such as APPBOSA[14] and APOQ, appear to utilize mainly family labor.[15] I further describe Dole's strained relationship with CEPIBO and union struggles in the next chapter. However, Peruvian FLO shipments via AgroFair (Netherlands) and Red Tomato (United States) are growing. In 2006 Peru's Network of Small Banana Producers (REPEBAN)[16] exported 120,000 boxes [2,400 tons] of organic FLO bananas at $8.43/box, nearly $2 above the price paid in Ecuador, totaling $27 million in value. This represented 60 percent of national organic production (*FreshPlaza*, June 19, 2007 [see chap 6, fn 21]).

Thus, while a notable amount of FLO-approved Andean production is done by smallholders, much of it—as with El Guabo, El Prieto, CEPIBO, and Colombian associations—is not. Some is grown and harvested in abysmal conditions and does not come from genuine cooperatives. While Fair Trade has benefited El Guabo and others, virtually all FLO-approved shipments are nonunion and freedom of association is unavailable to many workers. While the social premium has brought benefits, it has sometimes created divisions between members and workers and has been wielded by managers against unions.

Fair Trade and Plantations

Unionized plantations do possess freedom of association; however, they could substantially benefit from FOA scrutiny. FLO recognized this when it elaborated plantation criteria to expand sourcing and to meet market demand, as I noted in chapter 3. FLO encompassed FOA models such as Gariba Mustah's independent union at Volta River Estates in Ghana, where Gariba's co-workers enthusiastically endorsed Fair Trade as a way to voice their feelings about banana cultivation and sales (chap. 8). However, FLO failed to fully grapple with FOA elsewhere. By 2003 it had approved distribution (not production) through Dole, despite Dole's poor labor-rights record. TransFair, like other FLO organizations, began promoting bananas that included nonunion and Solidarista fruit from Costa Rica, and TransFairUSA discussed potential distribution arrangements with Chiquita. As northern demand grew, even OkéUSA, sought "a 6–7 percent gross margin to keep the business viable." It needed "much more volume," and hoped to "place longer orders with AgroFair so that when there is a shortage from one country, it could source from another" (interview with Rosenthal 2007). Yet FLO's plantation dealings created resentment among both farmers and unions.

The farmers believed FLO-approved plantations would replace them: "AgroFair's position is that Fair Trade was never just about small farmers, that as long as produce is Fair Trade certified, it is okay, but the producers are fighting back. They say that Fair Trade was set up for smallholders, and the farmers within want to keep companies out" (interview with Rosenthal 2007). CLAC, which represents most of the associations enumerated in table 12.1, remains ever vigilant about the capabilities of TNCs and large landholders to manipulate markets and undercut smallholder rights. CLAC's banana farmers in Costa Rica, the Dominican Republic, and Ecuador feared that if TNCs (and unions) entered Fair Trade production, they would consume more land for export and squeeze out cooperatives.[17]

The unions also expressed reservations. While COLSIBA had been an early Fair Trade endorser, it steadfastly opposed Dole and Chiquita's involvement with nonunion bananas. When COLSIBA publicized its opposition to Fair Trade in fall 2004, USLEAP also encouraged a dialogue between TransFair and COLSIBA to find a resolution. Unionists seeking freedom of association were therefore gratified when TransFair and COLSIBA reached an agreement in June 2005. TransFair promised that when sourcing from TNC plantations, it would ship only union-approved (not necessarily unionized) bananas. The two organizations shared "a common vision to organize, empower and raise the standard of living of farm workers . . . to collaborate further with the union movement . . . bearing in mind the interests of banana farm workers as well as those of small-scale producers." TransFair and COLSIBA committed to coordinate banana plantation inspections that included local union representatives, to work together on marketing Fair Trade union bananas, and to create a coordinating committee that evaluated progress biannually (TransFair/COLSIBA 2005). "It is clear that TransFair has this vision," stated German Zepeda. "It knows that half the bananas produced in Latin America are unionized. The banana TNCs have signed 20 contracts, all with us. Yet the 50 percent that do not have unions account for

80 percent of labor denunciations before the ILO. TransFair sees that if it wants to market a socially just banana, it must work with COLSIBA" (2007). In the accord's first phase, Chiquita was to produce and export bananas from a unionized Honduran farm and possibly complement supply with smallholder bananas.

Yet CLAC and other advocates of small-farmer production remained openly hostile to the TransFair/COLSIBA agreement, fearing a threat to the sustainable small-farmer system that they were committed to protect. The agreement also caused antagonism with other alternative trade NGOs, although TransFair had none on its board. Traditional supporters such as the Fair Trade Federation, Equal Exchange, Oxfam America, Lutheran Social Concerns, and Coop America voiced reservations.[18] By not consulting with its established CLAC partner, TransFair also "created a unilateral decision structure that was efficient, but weakened the system. In incorporating plantations, TransFair said 'trust us, we are on your side.' But the small producers responded, 'We have been run over for five hundred years. We want a voice.' It was the old colonial dynamic, with northerners going to help those in the South" (interview with Rosenthal 2007).

Labor supporters likewise queried whether the TransFair/COLSIBA accord truly protected all employees. They discovered a lack of clear guidelines, timeframes, and joint understandings. Along with guarantees of union freedom, they wanted checks and balances that resisted union corruption and the domination of particular patronage or ethnic group interests. For fledging unions, they required clearer premium-allocation mechanisms to avoid potentially damaging influences. They also sought postponement to designate procedures for certifying "union-approved" producers that had no collective-bargaining agreements. Consumers who paid a bonus for FLO bananas wanted assurances that it would reach both workers and smallholders in proportion to their productive role.

In late 2005, Hurricane Gamma devastated areas in Honduras, stimulating Chiquita to close certain fincas in apparent violation of the Global Labor Accord (see chap. 10). The hurricane ended TransFair's immediate plans to source union-approved Honduran bananas. Dole took advantage of the lacunae, courting TransFair to become its supplier for union-approved bananas. However, COLSIBA adamantly resisted Dole's advances, and to its great relief, TransFair followed the Coordination's wishes. After 2005, TransFair certified supplies of FairTrade bananas but did not immediately seek an expansion under union auspices.

Future Collaboration on Free Association

Despite TransFair's failure to consult with CLAC and the NGO community, its accord with COLSIBA represented an important first step toward a consumer-labor-smallholder alliance. It illustrated strategic opportunities, networks, and expanded identities. Yet freedom-of-association issues persisted among both suppliers and retailers.[19] FLO puzzled over how it could encourage cooperative leadership and worker empowerment via joint manager-worker bodies without undermining union organizing or independence. It pondered how it could reconcile its gradual, "developmental" approach when it appeared to conflict with the "best practice" of having a union. FLO legitimately insisted on certification independence, yet realized the value of union input (Zonneveld 2007).

In June 2007 IUF unions and FLO staff met in Bonn, Germany, under the sponsorship of the Friedrich Ebert Foundation (FLO/IUF 2007). Participants wrestled over defining worker empowerment in the face of low hired-worker wages, nonfunctioning complaint procedures, and management misuse of premiums. They debated communications, reconciling unions with manager/worker joint bodies, and the FLO-CERT inspection process. While a common understanding on freedom of association "had not yet been reached," FLO invited union members to serve on its standards and certification committees, and to become involved in inspector training. FLO and the IUF had acknowledged "an important step forward in the construction of such cooperation," forming a specific pilot in Costa Rica (FLO/IUF 2007). Toward year's end, a Fair Trade Foundation appraisal found FLO's vision still muddled; FLO-CERT certification had not improved worker pay, and conditions remained below conventional practice. Antiunion attitudes, inspector ambivalence, and failed complaint procedures were among the reasons. While also acknowledging a need for worker training, the appraisal urged a role for unions in improving audits, pressuring owners, and raising plantations to TNC standards (Williams 2007). In a related protest, the FLO-CERT union representative resigned.

FLO believed such characterizations were unfair, given its active engagement with unions: FLO-CERT claimed it had established a mechanism to deal with public complaints and a record of well-trained auditors who conducted off-site interviews to test worker representation. These steps involved unannounced inspections and warnings to plantations that prevented union access. In October 2007, FLO mandated specific freedom of association training for its central auditors (Perez 2008).

FLO's new leadership in 2008 reasserted its strong commitment that ILO principles be embedded in FLO-CERT implementation on the ground. It invited unions to develop a module for inspector training and expressed openness to union accompaniment during visitations. FLO and the IUF sought regular information exchange, with a union official serving as liaison as well as on FLO-CERT's Committee on International Certification.

CHAPTER 13

Dole
Reluctant Fairness

AS IT ENTERED the twenty-first century, the produce giant Dole had outmaneuvered its chief rival, Chiquita. It had fifty-seven thousand employees with sophisticated diversification in ninety countries. Company researchers had produced twenty kinds of bananas alone. However, despite Dole's recognition of United Farm Worker (UFW) strawberry workers and several unions in Honduras and Ghana, and despite its membership on the SAI board, labor cited Dole as a company consistently repressing union organizing and relocating without transfer agreements. During the middle of the decade, the company had thwarted unionization in Costa Rica and Guatemala while it expanded in nonunionized Cameroon and Brazil.[1] In Ecuador it turned a deaf ear to the FENACLE union as it sourced from as many as six hundred producers who paid their employees a scandalous $56/month, a third of what Central American field hands received (Perillo and Trejos 2000).[2]

In this chapter I consider union organizing challenges in Ecuador and Costa Rica. After Dole's noncooperation brought unheeded warnings, unions and supportive NGOs inaugurated a campaign titled "Behind the Smokescreen," designed to expose the company's antiworker practices, that brought limited results. At the same time, Dole moved agilely to exploit TransFair's boggled Fair Trade/union agreement with COLSIBA. According to Ron Oswald, IUF general secretary, Dole was "very interested in getting Fair Trade certification. It claims that all its workers have the right to be in a union, although very few of them do. It has used SAI certification to gain first preference among supermarkets, and probably would use Fair Trade certification the same way. To critics, it can say, 'Dole is in dialogue with the unions and we have SA8000'" (commentary to author, 2006).

Organizing in Ecuador

Union leaders have described how workers in Ecuador and Peru fear to speak about unions. "There is suspicion, not freedom of association

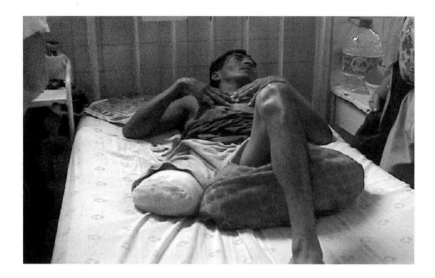

FIGURE 13.1. FENACLE worker Muaro Romero in Luis Vernaza Hospital, shot at close range by Noboa's men, Ecuador, 2002. (Photo by Jan Nimmo, Banana Link)

for bargaining and striking," emphasized German Zepeda of COLSIBA (2005). Nevertheless, the COLSIBA network sought opportunities, stimulated awareness, and campaigned to stem the growth of nonunion bananas. Owing to past repression, FENACLE's affiliate officially represented only 2,500 of Ecuador's 300,000 banana workers. In February 2002, 1,400 FENACLE workers targeted a job action at seven Bonita plantations owned by the nation's wealthiest businessman and presidential candidate, Alvaro Noboa. Although the workers gained legal recognition, Bonita stepped up firings. In May 2002 FENACLE inaugurated a broad strike. Alvaro Noboa immediately summoned thugs who terrorized and shot at the workers, sending nine to the hospital with gunshot wounds (fig. 13.1). FENACLE quickly evoked an international campaign, coordinated by USLEAP and EUROBAN, to persuade Noboa to negotiate a contract (USLEAP 2002a, 3). Promoters of Fair Trade bananas joined the campaign, and Human Rights Watch documented rights violations, child exploitation, and pesticide misuse.[3] Although the campaign did not persuade Noboa, the *New York Times* suggested that the candidate lost Ecuador's 2002 presidential election partly because of his response.[4] When Noboa ran again in 2006, FENACLE produced and widely circulated a video on workers' rights that featured their lack of adequate water and housing. Despite a first-round victory, the banana tycoon lost the vote again.[5]

However, Dole was not in the presidential race. It pushed workers and children as far as it could. Of the forty-five underage workers Human Rights Watch interviewed, thirty-two averaged eleven and a half years of age when they began plantation work (Human Rights Watch 2002). When they were contacted three years later, the same workers recounted that Dole had persisted in their mistreatment.[6] But if they organized, Dole would cut them out entirely.

In 2001, the TNC did nothing when its independent producer in Ecuador fired the general secretary of one of the country's few banana unions when he demanded contract enforcement. Dole's abuse of children proved persuasive: when workers formed unions on Dole supplier plantations Tara Del Sur, Francia, and Manuela in 2004, the owners dismissed enlistees. Managers fired 120 at Maria Elisa who attempted to organize. Pressure brought negotiations and eighty reinstatements but none to union leaders (*Dole, Behind the Smokescreen* 2006). In October 2005, when ninety-eight workers set up a union at the Josepha plantation that sourced both Dole and Noboa, the owner maneuvered recognition of a company union. For its part, Dole refused to recognize the union firings and illegal violations (USLEAP 2005, 5). The remaining workers struck Josepha for several months, requiring a centurion force of police and armed guards to evict them. The strikers finally gained a provisional agreement in January 2006, but not before the managers had locked out sixteen, twelve of them union leaders (*Dole, Behind the Smokescreen* 2006).[7]

Costa Rica

Dole's behavior presented special difficulties for the unions in Costa Rica, who concluded that the TNC was "probably the biggest single supporters of *solidarismo* (management-labor Solidarista committees)." In 2002, union workers at Acon, Dole's largest Costa Rican subcontractor, reported an increase in health and safety violations, especially aerial spraying over fieldworkers. Acon then mounted a major antiunion campaign at its Carrandi plantations. It first offered payoffs that included homes and jobs for family members. When this failed to deter worker complaints, Acon followed with threats, discrimination, and blacklistings. In 2005, the company ejected five union activists. After the union at the Las Perlas plantation gained new members, the solidarismo promoters threatened them and their families with wage reductions or dismissal, and blacklisting (*Dole, Behind the Smokescreen* 2006, 14).

Environmental Damage

Dole's ecological violations likewise angered union supporters, along with its advocacy of a tariff-only system reported in chapter 7, and its treatment of pineapple and flower workers.[8] In Costa Rica in 2003, Dole spilled three thousand liters of Bravo 72 fungicide in the water system near the Bataan airfield, killing thousands of fish in the Pacuare River. The TNC eventually agreed to pay $115,000 to restock the river and set up a pollution-prevention system. However, in 2005 toxic residues again appeared near Bataan and the Madre de Dios River close to Dole's Las Perlas farm (*Dole, Behind the Smokescreen* 2006, 21–22). Dole also was the company most heavily associated with the DBCP scandal in which 1,245 Nicaraguan workers who applied Nemagon and Fumazone became sterile. Dole responded that it used only "safe" products. When pressed, it acknowledged that the pesticide manufacturer, Dow, had warned about potential danger.[9] Nevertheless "it only applied DBCP 13 times in the open air" with proper instruction; when a local judge ordered the company to pay $804 million in compensation, Dole dismissed the judge as corrupt (Miller 2007).[10] The case came to U.S. courts, where every step was "marked by intrigue. Attorneys have been accused of fraud, and Nicaraguan workers have held hunger strikes. In addition, top Dole officials have met with Nicaraguan President Daniel Ortega to offer the possibility of jobs if the country made changes to its legal system that would make it more difficult to sue" (*Los Angeles Times*, July 20, 2007, B2). As explained in one trial document "a Dole official said it was 'well-nigh impossible' to implement some safety measures at the plantations." The U.S. jury awarded $3.2 million to six Nicaraguan families (*Los Angeles Times*, November 6, 2007, B1).

Dole and Fair Trade

About 2002, Dole created an organic products division and, despite NGO complaints, soon became a certified Fair Trade distributor. In addition to establishing a production unit of twelve hundred hectares in Honduras, the TNC marketed organics from the Dominican Republic, Ecuador, Colombia, and the Philippines. "In 2003, we signed with Fair Trade in Peru with 160 producers," boasted Dole Vice President Sylvain Cuperlier. "We now have registered production in 5 countries for the EU market. This is how we prevent the race to the bottom" (Cuperlier 2005).[11] Nevertheless, workers charged that Dole's "socially superior" bananas deceptively competed with union-produced fruit.[12]

This had likely been the case when Dole began importing organic bananas from Peru in 2001: it was paying only $1.90 per 18.14 kg box (fruit on plant), by far the lowest price in the hemisphere. When Dole sought to renew individual contracts at the end of 2002, Valle de Chira and Tumbres farmers found their need for more hand labor and organic fertilizer had substantially raised their costs, so they formed co-ops to negotiate. Dole's response "was the harshest. . . . They did not like us organizing. However, we obtained a price of $2.55 per box and they furnished the plastic bags" (CEPIBO 2006). Since costs remained high, six area associations then set up the Central Piurana de Asociaciones de Pequeños Productores de Banano Orgánico (Central of Peruvian Associations of Small Organic Banana Producers, CEPIBO) for more bargaining power.

By 2006, CEPIBO had thirteen hundred members, each owning a hectare of land. CEPIBO was selling more than fifty thousand tons of organic bananas for more than $12 million through three exporting companies, including COPDEBAN, a Dole subsidiary.[13] Dole was paying $3.00/box, the minimum Fair Trade price (fruit on plant) plus the $1 premium. Since FLO criteria encouraged producers to take part in the packing, CEPIBO sought COLSIBA's help to control the export process. "Dole tried to divide us using Fair Trade and offering to obtain certification for some but not all of us. . . . We thought it was unfair and we wrote to FLO and Dole." When FLO then considered certifying CEPIBO as a whole, Dole became "infuriated. They are cutting down on orders and punishing CEPIBO leaders, and, as the president of CEPIBO, I am one of them" (CEPIBO 2006).

Dole and its COPDEBAN affiliate subcontracted banana packing to other organizations. In the Valle de Chira, Peru, SERAGRO-SAC (a Dole-created labor contracting company) caused employee resentment in the way it supervised postharvest and palletizing operations. Twice, its workers tried to organize unions but were "sacked immediately," and BanaFair determined they were denied all rights (Fischer 2007b). Dole had acknowledged that its Peruvian Fair Trade employees may have been intimidated from speaking openly to monitors (Cuperlier 2005), but it took no observable action. CEPIBO described Dole's strategy in Peru as "very clever; the company breaks up organizations; it prevents unionizing but never publicly. Behind the smoke curtain, it denigrates any person who wants to become a union leader, explaining it is out of ambition or because she wants to have political power" (CEPIBO 2006). Ultimately after considerable effort, COLSIBA was able to assist SERAGRO workers to achieve representation through its new SITAG (Sindicatos de Trabajadores Agrícolas de Perú [Unions of Agricultural Workers of Peru]) federation and gain legal compensation for Sunday work.

Stakeholders' Campaign

By March 2006, the transnational union network had grown weary of Dole's intransigence. COLSIBA gave the company a five-month ultimatum to demonstrate its openness to organizing in Ecuador and Costa Rica. EUROBAN affiliates began contacting supermarkets. In May 2006, supporters charged Dole with failing to respect basic worker rights. COLSIBA, USLEAP, and six other organizations released *Dole, Behind the Smokescreen*, documenting violations.

In July 2006, Ecuador passed a very limited labor reform that restricted subcontracting to 50 percent of all workers on any particular plantation. Ultimately, FENACLE decided that some rules for subcontracted employees were better than none, and that it would pursue further changes under a future government. In September, the COLSIBA unions unanimously announced that Dole had failed to make any meaningful improvements in worker rights, either in Ecuador or elsewhere: with EUROBAN and CGT France (Dole-Europe is based in France), COLSIBA inaugurated a full-scale campaign. EUROBAN members circulated the *Smokescreen* booklet, BanaFair's presswork in Germany grabbed attention even from conservative news outlets; the CGT contacted Dignité, Dole's IUF affiliate in the Ivory Coast; 3F expanded its campaign in Danish supermarkets; Peuples Solidaires arranged a FENACLE tour; and USLEAP stepped up U.S. support, causing TransFair to put its dealings with Dole on hold.

Finally, Dole responded. In Norway, where it held 60 percent of the market, the company issued a seventy-page reply to campaign charges. To head off a major global movement, it expanded dialogue in Ecuador and signaled an accord with unions in Costa Rica. Dole's tentative agreement with Costa Rican unions in March 2007 pledged to ensure organizing access and respect for worker rights. The accord caught the transnational activist coalition off guard, since participants had agreed there would be no separate deals until final settlement. Nevertheless, COLSIBA viewed the Costa Rican accord as a test and insisted that the global campaign continue.

In Costa Rica itself, Dole had difficulty obtaining compliance from Acon and other local suppliers, especially at Collin Street and Anabel plantations, where Colombian owners and managers openly criticized unions. In Siquirres, Matina, the Valley of la Estrella, Guapiles, and Sarapiquí, Dole representatives dragooned workers to view an archived Chiquita video that lambasted labor leaders and threatened the consequences of a union choice. Valley of la Estrella's packing plants displayed anti-union posters. At the Porvenir finca, Solidarista personnel virtually blackballed an elected union representative. Overall, Dole stealthily encouraged solidarismo despite its union agreement.

In 2007, the newly elected Ecuadorian government credited its union backers with playing a crucial electoral role. It promised fresh legislation on subcontracting, wages, debt assistance, loans for the poor, and support for single mothers. Dole-Ecuador immediately hired a former ILO staff member to work on social responsibility. FENACLE and COLSIBA met with the company, hoping for an agreement similar to the Costa Rican accord, and Peuples Solidaires visited Ecuador in solidarity. Yet on Dole plantations, workers feared to join unions (COLSIBA 2007), but FENACLE successfully organized 600 workers without major interference in 2008.

In Peru, Dole/COPDEBAN began negotiations with SITAG. Dole's SERAGRO affiliate began Sunday pay in mid-October 2006 but had difficulty providing pay slips. On Guatemala's southern coast, conditions remained precarious. In August 2007, more than fifty NGOs wrote Dole, listing other ongoing violations.[14] The TNC replied that it was emphasizing Dole's corporate responsibility policy and Costa Rican agreement, which it promulgated at a forum it convened with the Banana Growers Association in Guatemala (Cuperlier 2007). Yet unions said paramilitaries moved freely on Dole's Guatemalan farms. In early 2008, when SITAG in Peru requested full compensation for Sunday work due since 2003, SERAGRO fired forty workers, stimulating an outcry to Dole from the COLSIBA transnational network, perplexing the company.[15]

In the twenty-first century, Dole and Noboa's treatment of banana workers and small producers in Ecuador and elsewhere became a rallying opportunity for all banana stakeholders. The FENACLE and Smokescreen campaigns helped workers, peasants, and Fair Traders increasingly realize a common agenda. In COLSIBA's view, by fall 2007 the situation on the ground did not "seem to have changed significantly . . . the company is now more reluctant than ever to try to find a solution in Ecuador, that we are witnessing an increase of anti-union activities in Costa Rica. . . . We consider [it] therefore necessary to continue the campaign" (COLSIBA 2007). EUROBAN members again petitioned Dole, with seventy-four organizations reiterating COLSIBA's points (EUROBAN 2007). Eying another opportunity, unions also urged that FLO adopt a more structured auditing procedure that assured labor rights, with Dole a key target. In hesitant steps, the TNC responded.

Summary

Challenges can test any budding banana alliance, and Dole's efforts to hold off unionization, yet co-opt Fair Trade, represented such a challenge. In this chapter I have explored how farmers, workers, and Fair Traders visualized opportunities, how they networked to confront both challenges, and how they generated fresh self-conceptualizations of themselves as an alliance in the process. While Dole saw a chance to upend Chiquita as the supplier of the supermarket banana of choice, COLSIBA and its EUROBAN allies mobilized. They established new network linkages to elicit support from African allies, Caribbean supporters, and northern consumer interests, convincing TransFair to delay certifying Dole as a U.S. distributor. Through the operation, they defined a "responsive consumer identity" as one that would take corporate manipulation into account.

CHAPTER 14

A Proposal for Fairer Trade

DEBATE ABOUT THE rules of trade reflects disputes over the optimal scale of banana production recounted in chapter 7. Changes to the European trade regime in 2006 risked further exacerbating tensions between small producers in ACP countries and TNC plantations in Latin America. In this chapter I summarize how small farmers, unions, and NGO EURO-BAN partners grasped opportunities, expanded networks, and strengthened identity to advocate five trade-fairness proposals: a multistakeholder forum, differentiated tariffs, recycled tariffs, legal-labor verification, and an international banana agreement.

Bananas After 2005

The WTO and Cotonou settlements of 2000 portended crucial adjustments to banana trade that favored a neoliberal trading system. They

1. phased out all quotas—Latin quotas by 2006, ACP quotas by 2008;
2. imposed a non-ACP tariff well above seventy-five euros per ton;
3. replaced quotas with Economic Partnership Agreements (EPAs) that the EU would negotiate with each ACP nation to help local development.

Many observers predicted that Europe's change from quotas to a tariff-only system would have a detrimental impact on both unionized bananas and smallholder production. Labor and farmer representatives both explained that this new emphasis on open trade undercut more balanced European policies that helped stabilize Windward and Caribbean livelihoods and democratic traditions.[1] WINFA, COLSIBA, and various EUROBAN partners unanimously articulated that a high tariff would have a negative impact upon workers and farmers. They insisted that the current tariff of seventy-five euros per ton remain in force "until a system that enables sustainable production can be introduced" (Parker and Harrison 2005). "In Costa Rica, for example, the banana area of Limón is the poorest of our political divisions. . . . We ask the tariff-only policy makers what they are going to do about this?" asserted union leader Gilberth Bermúdez (interview 2004). One producer said the change would force him out of the market and into cocaine production (Reuters, February 7, 2004). Higher fees for Latin bananas would motivate TNCs to increase exports from West Africa, with likely devastating effect on production from the Windward Islands, Jamaica, and Belize (Myers 2004). "The companies shift operations because fresh lands are available; they have a shorter shipping distance to Europe; the cost of production is even lower than in Ecuador; few unions exist, and the tariff encourages the move. If Europe can receive shipments from Africa, why should it maintain interest in the Caribbean, much less in Central America?" (interview with Bermúdez 2004).

Yet the TNCs divided over trade strategy. Chiquita insisted on the tariff of seventy-five euros per ton as "a common Latin America negotiating position. The changes ... offer us an opportunity!" exclaimed Craig Stephen, chief of Chiquita–Asia Pacific. "If we don't get it right, we won't be able to talk about improving labor conditions ... for the next generation. A high tariff will ... cripple the industry in [the Latin] region, and in the U.S." and encourage producers in the Cameroons and Ivory Coast (Stephen 2005).[2] However, Dole was willing to tolerate a tariff of between 113 and 130 euros/ton along with some ACP preferences. Del Monte remained neutral on tariffs but strongly favored elimination of the European licensing system through which Chiquita and Fyffes had gained advantage. "License holders get significant profits, and this leaves no funds for improving living conditions for workers," emphasized Marco Garcia, head of Guatemalan operations. "Under a free market, they can place the excess and better estimate the costs. The consumer will benefit" (Garcia 2005).[3]

In February 2005, without negotiation, the European Union announced a 230-euro tariff, outraging Chiquita: "The proposed tariff will damage the industry, adding $5 to a box of bananas, causing 95 percent of producers to sacrifice for the 5 percent who benefit. This is bad economics" (Stephen 2005). COLSIBA's German Zepeda asked why "TNCs would expand to other countries for only a two-year 'comparative advantage' [in 2008 when the Doha waiver for the Cotonou Agreement expired]?" He predicted "a huge negative impact on our workers and those in the ACP countries. Even the Canary Islands producers have joined us with their concerns, shifting their viewpoint 180°"(Zepeda 2005). Facing legal obstacles raised by TNCs, nations, and EUROBAN pressure, in January 2006 the EU Commission then imposed a somewhat reduced tariff of 176 euros/ton ($220 at that time). Concerned parties and governments filed additional appeals to the WTO.

EPAs

Meanwhile, Cotonou Agreement signatories had to establish new trading arrangements by January 2008 that ensured "the continued viability of banana export industries" (Agreement, Art. I, cited by Myers 2004, 121). The seventy-six ACP nations searched for a sensible EPA approach that would assure some equity among historically committed growers in the Windwards, Jamaica, and Belize; isolated plantations in the Philippines, Ghana, and Ivory Coast; organic cultivators in the Dominican Republic; and the emerging producers in Brazil and Cameroon. They faced an EU free-market approach that sought to compensate certain nations with special aid packages. WINFA Coordinator Renwick Rose was particularly worried. The regime had promised "that we should be no worse off than before, but it seems clear this could be overridden by the pressures of multinationals and supermarkets to get the cheapest possible bananas." Rose urged all governments to pressure the EU to honor its obligation, fearing that (even with EPAs and Fair Trade) the changes "could very well wipe out all the small farm industry in the Caribbean banana (and sugar) trade." He argued that Windward labor standards guaranteed a minimum wage "higher than what the multinationals pay in Latin America." (T and G Publications 2004).

The COLSIBA/WINFA/EUROBAN Response

The dire predictions of high-tariff opponents did not immediately occur, as table 1.1 reveals. Increased demand, bad weather in Ecuador, and savvy Fair Trade marketing combined to sustain banana sales. Costa Rica, which had lost market position in 2005, implemented severe cost-cutting measures to gain it back in 2006. The Ivory Coast, Ghana, the Dominican Republic, Guatemala's southern coast, and Brazil benefited from increased EU purchases. Exports from the Windward Islands remained stable in 2006 at around sixty-two thousand tons or 7.5–8 percent of the U.K. market (see table 9.1). Yet the low production in Ecuador created wide swings in prices, from an amazing $11/box in January 2006, down to $1 in August, $2 below the official limit. While Dole did well, Noboa and Chiquita/Reybanpac lost sales. Some small producers teetered on bankruptcy. "The export companies are therefore able to buy more and more of our land at rock-bottom prices. We are desperate," lamented one El Oro farmer (COLSIBA, EUROBAN, and WINFA 2006). Overall, Ecuador lost 3 percent of the EU market, and Colombia nearly 2 percent, for an $80 million loss. Following Chiquita, which had divested itself of its Banadex affiliate two years earlier, Dole sold its Colombia marketing firm, Proban. Two national companies, Uniban and Banacol, took control of over 80 percent of Colombian exports. Luc Hellebuyck, of Bananic International, suggested that while the four largest TNCs increased sales by 5 percent compared with the same six-month period in 2005, their profits dropped by 75 percent, and their market share fell by 1 percent, "indicating that other commercial players have benefited more from the EU market liberalization" (EUROBAN 2006).

Small farmers in the Windwards were able to maintain a relatively steady volume of banana exports in 2005 and 2006 for two reasons: First, FLO certification stopped the decline in island banana farms and even helped establish a small plantation where WINFA encouraged an independent trade union. Following Fair Trade Foundation efforts, the Sainsbury chain in the United Kingdom committed to purchase 75–80 percent of St. Lucia/Dominica bananas. WIBDECO gained FT certification and agreed to contract exclusively through WINFA, placing farmers and workers in control (!). As Fair Trade expanded, Windward farms hosted more visits from researchers, media, and supermarket representatives. Over all, the Windwards increased their Fair Trade/export ratio from 29 percent in the first quarter of 2005 to 90 percent in 2007, even though Hurricane Dean temporarily reduced sales.

The second reason for success was crucial but transitory: although the islands still held a guaranteed 40 percent of ACP banana sales, the formula would disappear in 2008. After this date, the high tariff would still protect ACP countries, but no set-asides would assure market output for specific islands. The prime minister of Dominica warned that this would stimulate production in other ACP areas and "inevitably result in Caribbean suppliers being forced to pay the duty of 176 euros/ton," squeezing Windward farmers (EUROBAN 2006, 3). One study estimated that the end of the guarantee would also cost Latin workers 115,000 jobs. "We find ourselves really being confronted by the power of the United States, the power of the multinationals and the power of the Latin American producers," concluded Dominica's Minister of Foreign Affairs and Foreign Trade Charles Savarin (*Jamaica Gleaner*, November 9, 2006).

EU officials touted Economic Partnership Agreements with concomitant agricultural assistance as the solution to specific national economic problems. The EPAs required reciprocal market access and protected intellectual property, but they fell short in detailing public benefits. Five U.K. development agencies scathingly described them as a "Partnership Under Pressure." They charged that the European Commission had forced "onto the table issues rejected by the WTO . . . linking future development assistance to the trade concessions . . . [that] run into $ billions" (U.K. Food Group 2007).[4] Trade unions issued their own report asking to exempt bananas from EPAs, along with sugar and rice.

WINFA likewise struggled diligently with other ACP farmer networks as well as with small producers from Cuba to delay the EPAs.[5] With the Caribbean Trade Partnership, to maintain quotas, WINFA lobbied WIBDECO to avoid intermediate relations with TNCs. WIBDECO gained direct supermarket outlets with Tesco, Sainsbury, Waitrose, and others, while WINFA, assisted by Oxfam, mobilized major rallies on each island, culminating on June 3, 2007, with the Mt. Bentick Declaration opposing immediate EPAs. WINFA conducted a broad EPA educational campaign via a score of farmer clusters on each island. It rallied thousands of small growers, association leaders, and political officials to challenge CARICOM officials (representing the Dominican Republic, Belize, Surinam, Jamaica, and the Windwards) to urge EPA postponement and demand protections beyond a differentiated tariff. Unions in Ghana and the Ivory Coast put similar pressure on African governments.

Political Responses

In February 2007, Ecuador formally petitioned the WTO's dispute-settlement panel to investigate EU violations of the 2000 trade-regime agreement (chap. 7). The EU had offered some Latin America countries a "peace clause," whereby it would lower the tariff to eighty-eight euros per ton over six years (Costa Rica had accepted after recovering from 2005 losses, although the WTO had yet to approve). Ecuador was chosen as a lead petitioner to oppose high tariffs and ACP preferences. Despite improved sales in 2005, Ecuador had endured a 4 percent drop in sales in 2006 (see table 1.1). President Correa accused the EU of having "no will for fair trade" (*FreshPlaza*, July 19, 2007). The United States followed with its own WTO filing against ACP advantages. Colombia and Panama (with Honduran and Nicaraguan backing) also claimed unfair competition since producers with similar expenses received differential treatment at EU borders. Chiquita encouraged the countries' legal claims, saying the 176-euro tariff had cost the company an additional $75 million in 2006 (Miller 2007). ACP nations also opposed the Latin peace clause as a move toward a lower tariff and against any possible quota arrangements—in other words, an assault on small farms.

In late 2007 the WTO dispute panel ruled in Ecuador's favor, rejecting ACP quotas and duty-free treatment. The EU leveraged the finding to divide the Caribbean Forum of African, Caribbean and Pacific States (CARIFORUM) and wrest WTO-compliant EPA agreements from individual countries. Although none had yet signed an EPA officially, in early 2008, the EU opened its market to duty-free, quota-free access for CARIFORUM goods. About the same time however, the WTO announced that the 176 euro/ton tariff on non-ACP bananas was excessive, opening the door to national trade sanctions. This could arguably aid Ecuador, but it would also endanger small island producers.[6] By July, the CARIFORUM nations had yet to put the official pen to any EPAs.

Peace at Last?

On the twenty-eighth day of the same month, the EU, United States, and Latin and ACP nations finally accepted "The Geneva Agreement on Trade in Bananas," which would reduce the 176-euro tariff by 35 percent over a seven-year period. The accord would end all previous banana-tariff disputes and commit to resolve future complaints within the WTO. ACP nations would receive an aid package to help restructure their industries. Unfortunately, at the end of July the WTO's Doha round

TABLE 14.1. EUROBAN Proposals to Address Banana Trade Issues

Proposal	Anticipated Outcome
Multistakeholder forum	Dialogue between civil society, WTO
Differentiated tariffs	Nations observing criteria pay less
Recycled tariffs	Producers share tariff benefits
Public verification of labor rule	Restored role for legal mechanisms
International banana agreement	Adds social, environmental clauses

Source: Drawn from Lombana and Parker 2006.

of trade negotiations collapsed. The EU scuttled the Geneva agreement. By October, CARIFORUM banana states had all signed EPAs.

Coalition Stakeholder Recommendations

EUROBAN confronted such divisive political manipulation with five proposals for a sustainable banana-trading policy that would protect future interests of both small producers and workers (listed in table 14.1).

EUROBAN'S proposed forum would inaugurate discussions among national WTO representatives, unions, and NGOs, including groups like CALC/Via Campesina, ILO, FAO, WINFA, IUF, GAWU, and COLSIBA, to develop the principles for a coherent and fair international banana-trade policy. Southern states would convene the forum and establish the agenda, which would include, for example, the African plan for supply management.

Differential tariffs would be imposed according to social, environmental, and/or economic criteria. The tariffs would require mechanisms for weighting, standardizing, and ensuring criteria compliance, including what constitutes fair trade. Even though bananas have so far been excluded from the European Union's trading program known as Generalized System of Preferences Plus (GSP+),[7] EUROBAN members indicated the EPA "quota replacements" might become part of such an approach. They called for "a system of differentiated tariffs within the new EPAs so that only bananas produced according to some kind of sustainability criteria [would] enjoy duty-free access." This system would give ACP countries an incentive to improve conditions, and fair-trade producers preferential access (Lombana and Parker 2006, 5).

The recycling proposal would return 25 percent of any imposed tariff to local producers. The funds would be restricted to sustainable and equitable banana production, that is, to fund research, implement labor standards, bolster fair-trade programs, improve small-producer productivity, and so on. The recycling could be incorporated into the current EU import regime without major modifications.[8]

Finally, EUROBAN analysts suggested possible public mechanisms for verifying labor and environmental rights. These could involve potential clauses on labor and environment in a European Union–Central America trade agreement that would declare bananas as a sensitive or tropically traded product. One specified outcome could be the restoration and funding of state labor-inspection systems as called for in ILO Convention 81.

The EUROBAN network and its allies continue to pursue these proposals. Despite their demands, ninety NGOs were unable to prevent EU commissioners from postponing EPA finalizations, but they still hoped to improve government transparency and democracy as preconditions for any agreement and its implementation:

> As the EPA negotiations move towards their climax, we are deeply concerned to ensure that the livelihoods and future of millions of poor people are placed at the core of the negotiations, and that these priorities are not sacrificed . . . [A]ll the signs emanating from the negotiations suggest that these commitments are . . . antithetical to development. . . . Europe has achieved regional integration by protecting and promoting its regional market, and supporting producers in agriculture and industry to become competitive. Yet . . . ACP regions . . . are being asked to open their regional markets to the EU before their producers and regional markets have had the opportunity to mature. . . . We recall the EU Council's commitment of 15th May 2007 to "fully respect the right of all ACP States and regions to determine the best policies for their development."[9]

With tariffs affecting Latin bananas in the European market (and more bananas going to the United States), Fair Trade and individual producers like Heriberto Custodio and George DeFreitas, as well as union producers like Jiménez Guerra, Auria Vargas, and Ramón Barrantes, face difficult decisions. With a tariff-only program, Fair Trade could also be jeopardized, making a banana alliance all the more important.

Summary

Trade challenges test any budding banana alliance. Despite the coalition's loss to a tariff-only trade program and the imposition of exploitative Economic Partnership Agreements, the unions and WINFA sustained their lobbying for a just alternative. After documenting the actual impact of the new trade regime, EUROBAN network members offered a five-part plan to recapture a just trade balance. The FAO and UNCTAD (United Nations Commission on Trade and Development) joined their call for a multistakeholder forum in 2009. Four hundred thirty-eight deputies to the European Parliament signed a declaration to control corporate and supermarket buying-power abuse. Whether or not the proposals gain traction, they do represent a coalescence of group interactions, the subject of my final chapter.

CHAPTER 15

Toward a Sustainable Banana Alliance

FAIR BANANAS FOR small farmers, workers, and consumers have been my focus in this study. I tested the hypothesis that the three groups could achieve common understanding and pursue joint action. In the past independent farmers would negotiate prices on their own; workers would battle huge corporations; and consumers might seek to patronize responsible growers. But farmers and workers were from the South and most consumers from the North, so I asked whether these separate interests could unite in a social movement.

Movement theory urged us to consider banana structures, networks, and identities. We looked for structural opportunities all along the food chain, such as when corporations imposed cultivation requirements; when pesticide contaminations limited banana output; or when hurricanes, shipping technologies, and/or free-trade regulations encouraged competition between independent producers and TNCs. We considered social networks as they lobbied for equitable peasant livelihoods and state regulations. We watched as small farmers formed associations, unions coordinated their actions regionally, supporters publicized conditions, and fair-trade groups set common standards.

Finally, in testing whether some type of common identity aided collaboration, we discovered a self-definition around fairness emerging among the groups. Although the identity was only partial, I demonstrated a certain agreement: small farmers came to appreciate their unique ecological role; workers realized the value of small farmers, as well as of their own links to consumers; and even TNCs, supermarkets, and national producers encouraged buyers of "quality" bananas to share a common project with banana workers—an important step in building consumer-producer awareness, no matter its intent. We saw that Fair Trade offered an especially persuasive identity for fairness; yet partners warned that Fair Trade must guard this identity from being subverted by insufficient consultation (and nonunion bananas). Ultimately, Fair Trade organizations publicly emphasized the importance of both farmers and unionized workers, and consumers consciously selected fruit on the basis of loyalties to both. Some organizations even lobbied to equalize trade regulations. As COLSIBA, WINFA, and EUROBAN informed the public about Chiquita's fudging and Dole's smokescreen of "responsibility," they manifested a unity for ecology and equity. Dissipating the concerns articulated by Fridell and others, they put forward a concrete plan of action for sustainable development.[1]

Agenda for Action

The coalition plan emerged from the Second International Banana Conference, "Preventing a Race to the Bottom," held in Belgium in April 2005 (IBCII). The final statement described the relevant structures, networks, and identities involved. It spoke of "structural over-production in the international market, coupled with the accelerated search for a cheap banana by big retailers" that put workers, producers, and marketing companies in "permanent crisis." It called for a full appraisal "of the economic, social and environmental impacts of different tariffication and supply management scenarios." Failing to delay a tariff-only system for the European Union, it nevertheless demanded "a comprehensive evaluation of the existing banana regime and its impact on poverty, income, wages, the environment, levels of development in all exporting countries, incorporating a gender analysis." Such steps were absolutely required for assessing any newly proposed bilateral Economic Partnership Agreements (EPAs). In addition to enforcing ILO Conventions on freedom of association, the conference proposed "that governments and international institutions promote programs, which benefit small producers through stable or increased markets for their produce, including through fair trade." Within the WTO, this would make "policies governing the banana trade be consistent with the concept of sustainable production" (IBCII 2005b).

The IBCII agenda demanded a unified network identity around six challenges: fair producer prices, union monitoring, consumer education, retailer responsibility, appropriate producer technology, and Fair Trade/FLO coordination. In fact, conference members endorsed a Fair Trade label that would represent union and smallholder interests. While acknowledging that the pursuit of these challenges would require constant monitoring from alliance participants, attendees agreed to coordinate consumer education and outreach to convince merchants to display and promote Fair Trade/union bananas. The group outlined potential contributions of each stakeholder, even those of unionized plantations that had not yet gained FLO approval. The action agenda represented a significant historical step; nevertheless, implementing its recommendations would require a vigilant effort by Fair Trade promoters, banana unions, and small banana cooperatives, one they would have to overcome divergent realities to sustain.

Structural Opportunities and Challenges

Across the globe, bananas are grown and exported almost exclusively by large fruit-marketing companies, and more recently, by supermarkets that source from two export systems: corporate-controlled plantations and small-producer cooperatives. In both, workers and small producers face common opportunities for resistance, but they also experience important differences that we explored. WINFA, CLAC, and other small producers are in some measure "owners," even if their ownership involves only produce, not land. They are structurally linked to what they grow in a different way than plantation workers are. They assume certain risks brought by climate and spoilage that field or packing hands do not assume. In theory, they have more freedom to act. They can personally decide what and when they will plant, harvest, pack, and transport. The workers hired by TNC banana companies are in a different structural place. Managers dictate the preparation, planting, harvesting, and packing regimen; specify irrigation channeling and pesticide applications; and pay wages according to daily or hourly tasks. While some companies negotiate, most dictate when workers begin in the morning and retire in the evening to assure compliance in all these tasks.

As we saw, however, many of the theorized structural differences between peasants and workers largely dissipate in the real world of TNC and supermarket control. Anton Bowman, of WINFA, described how farmers were called to produce more on the same land, with more fertilizer and pesticides, for "better cosmetic quality. At the same time the price for bananas was going down . . ." (interview 2005). Caribbean small producers often found their daily activities as dictated by Geest to be much closer to what large companies required of their plantation workers. We likewise witnessed how, owing to their resistance to United Fruit/Chiquita and other TNCs, plantation workers finally gained a structural advantage by achieving independence via relatively strong unions that provided housing, benefits, and transportation. Nevertheless, beginning in the 1980s, unionized workers increasingly found their standing jeopardized by changes in global trade arrangements, climatic conditions, and competition from nonunionized producers. In consequence, union affiliates also found their struggles mirroring those of smallholders.

Events in the late 1990s such as Hurricane Mitch further transformed these structural similarities. Europe had begun to shift from a quota to a free-market system. Tens of thousands of Windward farmers were forced to abandon production. As independent Latin producers attained market advantages, TNC growers gradually shifted to Africa and Brazil, jeopardizing thirty thousand unionized jobs. We saw how WINFA, COLSIBA, and CLAC struggled to transform differences into opportunities, the first two seeking a common resistance.

Fair Trade as a Structural Opportunity

Consumer consciousness about fairness arrived as a deus ex machina structural opportunity for the banana alliance, especially for those concerned about the race to the bottom in production and trade. Responding to public enthusiasm for the indigenous proposal of fair-trade coffee, northern national labelers stepped up certifications. When these organizations turned their attention to other products, they encountered TNC plantation production and the mistreatment of workers on larger estates. We witnessed how Fair Labeling Organizations International (FLO) standardized criteria for *both* smallholders and plantations, emphasizing local control, democratic principles, and "core labor standards" of the International Labor Organization (ILO).

Our findings reflected the different traditions of small-farm and union decision making. Each holder had his or her own way of planting. When FLO offered the social premium, it was natural for each group to debate how it would be collectively allocated.[2] In the Windwards and the Dominican Republic, we saw how a project would usually not go forward unless the majority of peasant participants agreed on expenditures. But unions had less familiarity with a democratic work process, or with open voting for allocation of a premium. Even though their members discussed and ratified contract proposals, unions responded to company-imposed rules reactively; they rarely formulated policy. Given their milieu of patriarchal traditions and loyal relationships, they had to adjust to distribute premiums democratically. I conclude that a unified approach would require that unions receive special training in the impartial apportionment of fair-trade opportunities. Two structural elements are also essential: a stakeholder arbitration mechanism to resolve union/FLO-CERT differences in particular circumstances; and an assessment tool to evaluate the functionality of the social premium in assuring labor rights (perhaps making it part of contract negotiations).

Mobilizing Networks

Networks enmeshed in the banana alliance included small-farm activists, engaged union leaders, labor support groups, fair-trade proponents, and responsive consumers. On the producer side, we saw how movement organizations like WINFA, COLSIBA, and the IUF had to first shore up their own network organizations, for example, over a common approach to Chiquita. I also described how they acted through EUROBAN and USLEAP to actively bridge differences and build strategic relationships. For their part, EUROBAN and USLEAP networked to convene smallholders, unions, and NGOs for International Banana Conferences in 1998 and 2005, and to support specific campaigns around Del Monte, Ecuador, Dole, and trade issues. We considered divisions among Fair Trade proponents, contrasting TransFair's broad marketing approach with the small-scale emphasis of other certifiers. In efforts at unity, EUROBAN reached out to FLO, USLEAP communicated with TransFair, and FLO and the IUF opened channels with each other, portending other Fair Trade links.

WINFA and CLAC

As networks, WINFA and CLAC advocated for all their members. With banana exports coming to constitute the bulk of Windward foreign exchange, we saw how WINFA courageously struggled to prevent the squeeze on family farms. Emphasizing sustainable methods, it registered thousands of small farmers as Fair Trade producers. While they encountered difficulties locating market outlets and equitable prices, they attained environmental training and community benefits. WINFA producers acknowledged Fair Trade as "an important step towards ensuring justice and a fairer trading system in the entire world" (Renwick Rose, quoted in T and G Publications 2004). The network explored markets for other products, forged alliances among other farmer organizations, and pressured local officials to take forceful positions on behalf of vulnerable producers. These actions had some influence on CLAC. Although it opposed a Fair Trade designation for corporate-produced bananas, CLAC warmed to trade union claims. As difficulties mounted, CLAC, Via Campesina, and WINFA positioned themselves to mobilize for a unified trade policy.

COLSIBA

We witnessed how workers affected by the corporate onslaught created the Coordination of Latin American Banana Unions (COLSIBA) as a network for sharing contacts and strategies. COLSIBA promoted an ecological agenda to improve local work conditions. It gained a remarkable worker-rights accord with Chiquita that opened all plantations to union organizing. At times its leadership could be sporadic and uncommunicative, oriented too much toward northern NGOs rather than toward local organizational development. Nevertheless, COLSIBA cultivated a solid relationship with WINFA and other Latin smallholders. Its affiliated workers in Sixaola, Sarapiquí, and Siquierres, Costa Rica; La Lima, Honduras; Izabal, Guatemala, and elsewhere were knowledgeable and open to coalitional work. One member union shared offices with a Costa Rican peasant association; another organized both smallholders and workers in Ecuador. In certain nations, COLSIBA persuaded co-ops and mid-sized farmers to hire organized workers.

The IUF

As a transnational labor network, the Geneva-based International Union of Food and Allied Workers (IUF) played an indispensable role in achieving the Chiquita global-rights agreement and in aiding the banana campaign in Ecuador. As an active member of EUROBAN, the IUF, along with COLSIBA and others, gave major impetus to a Fair Trade label that would highlight labor rights. The IUF saw a "direct link between union strength, consumer safety and public health and the environment. Where workers are denied the right to form a trade union and exercise oversight over health and safety, there can be no guarantee of safe food for consumers" (Longley 2005). Sometimes the IUF had difficulty coordinating regional affiliates, especially in the Americas. Nevertheless, Alan Spaulding, director of global strategies for the IUF's UFCW affiliate and member of the SAI and USLEAP boards, envisioned great potential for a Fair Trade/ union alliance: "If done properly, it is a project that supermarket workers can boost with enthusiasm" (Spaulding 2005). In addition to British, French, Irish, and Scandinavian affiliates, we also found probable North American labor participation and Fair Trade support with the Canadian Auto Workers (who also represent supermarket workers), the International Brotherhood of Teamsters (that includes many food-processing workers), the United Farm Workers, and the Farm Labor Organizing Committee. In its African Land and Freedom program, IUF affiliates worked with smallholders to negotiate improved prices and input arrangements. While GAWU encountered challenges in "the operationalization of its programmes" (Hurst 2007, 83), it anticipated future collaboration over fair employment and worker health.

Solidarity Center

The American Center for International Labor Solidarity remains an important support network for banana organizing. The center has exposed how the global economy harms rank-and-file interests through projects that emphasize worker rights, women's empowerment, and child dignity. Funding cutbacks from the U.S. Government and the National Endowment for Democracy have curtailed the center's banana organizing support, but in their work addressing violence and labor-rights abuse, center staff offers crucial assistance to a union/Fair Trade alliance.[3]

NGO Labor Supporters

We saw how transnational labor NGO support groups became another mobilizing network for informing activists and consumers. As network bridgers, they brought together island and COLSIBA representatives, forging a common impetus toward sensible trade policy. EUROBAN member groups like Banana Link, in the United Kingdom, and Bana-Fair, in Germany, helped monitor the Chiquita accord and protest the tariff-only system. Other EUROBAN members assisted in publicizing antiunion producers, advising on market considerations, and facilitating consumer pressure on the Ecuadorian government.

One important example of such an NGO in the United States was USLEAP, which cultivated trust among Latin workers and facilitated union contacts in the North.[4] USLEAP collaborated with the Development Gap and LAWG to sponsor workshops for Central American trade unionists and small agricultural producers on worker-rights provisions in international trade agreements. In a review process that involved the U.S. State Department, the AFL-CIO, Change to Win, industry associations, grassroots and advocacy groups, and Latin American rank and file, USLEAP helped muster cooperation from the U.S. Trade Representative to gain greater space for union organizing and some wage improvements.[5] USLEAP also facilitated North American religious, human rights, and trade-union partnerships with Latin labor unions and workers.[6] It sponsored demonstrations at malls and stores across the United States that clamored about worker issues in Nicaragua, Honduras, Mexico, Colombia, Ecuador, El Salvador, and Guatemala.[7] Along with the IUF, COLSIBA, and EUROBAN, the organization also played an important role in the Chiquita accord and the Dole campaign. On the USLEAP board is the former director of STITCH, Beth Myers, who led STITCH to establish programs that encouraged women banana workers in their own leadership development.

Other network organizations also assisted, including the International Labor Rights Fund, the Latin American Working Group, the Washington Office on Latin America, and the Labor Council on Latin American Advancement. Most participated in the North American Preparatory Conference on Bananas in 2005. Latin workers subsequently joined North American workers to protest Wal-Mart's 25 percent banana-price reductions in the United Kingdom (Banana Link 2007).

Fair Trade Networks

It is important to reemphasize how Fair Trade Labeling Organizations International traced its origins *from* a network of peasant producers in

Oaxaca, Mexico, that inspired northern NGOs like Solidaridad, Oxfam, TransFair, and Oké. FLO added bananas to its list of covered products in the mid-1990s because southern producers felt crushed by long-term price declines. While other alternative traders like BanaFair had preceded it, we discovered that FLO's entry inspired a broad mobilizing network of smallholders and northern advocates. Oxfam affiliates and others organized promotional events in schools and communities, and helped local cities and towns become Fair Trade sites where bananas and other FT products were widely available. Their actions convinced some unions to urge employers to handle or purchase FT products. In the United States, TransFair's retail contacts began requesting Fair Trade bananas. Churches, synagogues, temples, environmental groups, peace groups, political organizations, student associations, and consumer networks promoted Fair Trade products. While FT bananas still represented a market "niche," their expansion nourished a broader self-definition among promoters.

Reframing Identity

Identity proved to be the most difficult of the three requirements for a common movement, but I found indications of its presence:

1. Unions adopted a more unified position on fairness via the COLSIBA network, the Chiquita accord, the Dole campaign, and Fair Trade.
2. Small farmers such as those in WINFA adopted a fair-trade orientation, despite expressing their reservations.
3. FLO and Fair Trade organizations grappled with worker conditions and union rights at local, distributional, and international levels. Consumers also expanded awareness.
4. All groups grew more conscious, some poignantly, of land use and scale of production and distribution, although coherency around policy issues was less uniform.
5. Achieving partially shared identities facilitated bridging between the groups. Rejecting dogmatically defined class definitions, WINFA, COLSIBA, and FLO affiliates sought an approach that favored banana workers and consumers. Fair Traders and unions inaugurated discussions on how to reserve market segments for each type of producer.

Nevertheless, despite the application of FLO social-development criteria and ILO conventions, and despite cooperation with EUROBAN, most plantation bananas that FLO and other fair-trade groups imported into Europe were still not produced with union labor. Worker co-ops continued hiring additional laborers at below-minimum wages with little heed to union opportunities. COLSIBA repeatedly advised against Fair Trade fruit produced by workers who lacked contracts. "Under the guise of being more socially responsible, these bananas subvert union bananas," warned German Zepeda of COLSIBA. "Marketing these 'socially just' bananas may actually violate FLO standards by undercutting worker rights" (Zepeda 2005). Likewise, FLO rigidly applied environmental rules but was inattentive to the impact of inflation. Smallholders committed to Fair Trade could not fully count on the sale of their banana output, jeopardizing their livelihoods. There was a risk that Fridell's warning that Fair Trade practice imitated capitalist trading would be fulfilled, despite Barratt Brown's vision of an alternate system that linked southern producers with northern consumers (Barratt Brown 1993).

To compensate, identity coherence demanded crucial stakeholder adjustments. Besides increasing the representation of women in its directorate, COLSIBA had to engage affiliates in bona fide discussions about FT possibilities. When TransFair USA agreed to collaborate with COLSIBA, it had to be scrupulous about sourcing plantation bananas only from union-approved sources. FLO sent representatives to EUROBAN and convened with the IUF. While inconsistent in its practice, FLO did promise to await a green light from a country's independent unions (and smallholder groups) before FLO-CERT conveyed its certification. Prior to their promotion of the social premium, FLO-CERT representatives had to become knowledgeable about how smallholders and unions saw the world. They also agreed to reevaluate registered plantations that local trade unions listed as problematic.

FLO partners like the Fair Trade Foundation and Oxfam have been more scrupulous than many other organizations about respecting bona fide unionization as a measure of free association. Oxfam's moral mission lent credibility to the fair trade/labor/consumer cause because the organization also realized the importance of challenging the overall structure of inequitable trade. Oxfam encouraged consumers to lobby for trade rule changes that opened markets to third-world products and granted greater control to local producers. Oké's Jonathan Rosenthal looked to "a respected NGO like Oxfam" to "build trust," and "convene a forum that unions would attend" (interview 2007).

While alliance members have made identity adjustments, admittedly, preserving even this partial identity requires considerable attention. For example, as identities are bridged to produce FLO-certified fruit, COL-SIBA will have the daunting assignment of improving its own coordination and communications capacities, devoting more attention to local organizing than to European fund-raising. We saw how COLSIBA wrestled with breakaway or restive factions in Colombia, Guatemala, and Costa Rica. Organizing nonunion sites will require additional patience and care. So will involvement in on-site monitoring, premium allocation, relations with CLAC producers, and other applications of FLO criteria. Despite its Fair Trade enhancements, WINFA likewise faces logistical challenges with island groups.

A Costa Rican Model

One local banana coalition exemplifies the convergence of opportunity, network, and alliance identity. Foro Emaús is a collection of thirty grassroots groups in Costa Rica that support healthy, fair, and sustainable banana production. The Foro has skillfully woven disparate organizations together to share a common agenda. Shortly after its founding in September 1992, the Foro mobilized twenty-five hundred people for a "March for Life and Human Rights," urging legislators to regulate the environment and reform labor laws applied in the banana-growing areas. In subsequent years, the Foro, in collaboration with local banana unions, organized protests, high-profile tours, and national conferences on social and environmental banana standards and the Ethical Trade Initiative (Foro Emaús 1999a). The alliance also documented the difficulties that oversupply caused for both the industry and the unions (Foro Emaús 1999b; Thiele 2000). The Foro and unions jointly addressed specific attacks against workers (Vargas and Hermosilla 2000; Foro Emaús 2003; Hermosilla 2000). It conducted professional studies on pesticides, several in conjunction with COLSIBA and the IUF (Foro Emaús 2001; Foro Emaús and the International Union of Foodworkers 2001).

While relations between unions, environmentalists, and alternative trade advocates in the Foro have not always been smooth, the interaction helped labor groups expand their ecological understanding of fairer trade. Costa Rican unions often hosted the Foro's monthly assembly that rotated among various banana-producing areas. The Foro also arranged visits for international environmental groups, and promulgated *its own* set of social/environmental standards, pesticide buffers, and so on, which the unions helped develop (Foro Emaús 1998b). Workers and environmentalists in the Foro network forced corporations to better their social and environmental practices (see Foro Emaús 1998a, 2003). In one indication of the Foro's success, Del Monte cleaned up four hundred truckloads of waste and pesticide-laden plastic bags from the Baribilla River and lined the banks with tree plantings (Arguedas Mora 2002). Within Costa Rica, the Foro implemented a fair alliance similar to the one EUROBAN, USLEAP, the IUF, COLSIBA, FLO, Banana Link, BanaFair, TransFair, Oxfam, AgroFair, and other supporting groups sought to establish on a global scale.

Meeting the Action Agenda

A successful coordination of a banana alliance therefore joins the three strands of the global banana movement, tying the struggling efforts of banana farmers like George DeFreitas, Nioka Abbott, Juan Quenteña, and Heriberto Custodio with the efforts of workers like Gariba Mustah, Jiménez Guerra, Auria Vargas, and Ramón Barrantes. Yet as I have shown in previous pages, the coalition faces major challenges in assuring the Action Plan's six goals of fair producer prices, union monitoring, consumer education, retailer responsibility, appropriate producer technology, and Fair Trade/FLO coordination, along with improvements in trade rules, stakeholder arbitration, and social-premium dispersion. Price is critical for Fair Trade producers. Consumers are willing to pay more for Fair Trade bananas, but not substantially more, especially in times of economic turmoil and high energy costs; yet the price must be a fair one.[8] This fairness is what justifies the Fair Trade/union coalition's promotion of a rational banana-trade regime that proportionately assures adequate producer compensation. EUROBAN's proposal for assigning differential tariffs to socially and environmentally responsive producers, and its other four points for achieving balanced trade are illustrations of how a fair-trade regime might work (chap. 14; Parker and Harrison 2005; see also Preville 2005).

For FLO plantation bananas to be union approved, a first step necessarily involves clarifying the meaning of union approval in a way that grants power to both smallholders and union affiliates. FLO must be confident that union approval guarantees a process independent of special interests or personalities, and that it commits a set-aside percentage for smallholders. COLSIBA, African, and Asian unions must be assured that wherever possible, the remaining bananas are produced, grown, and transported by workers who freely organize and bargain, and then are promoted in a way that farmers, NGOs, and consumers can comprehend. In cases of conflict, the two parties would invoke their arbitration process.

But a lasting alliance requires a further step. Smallholders seeking self-sufficiency and cultural sustainability need Fair Traders to publicize the technological difficulties farmers face. Alliance stakeholders may ask retailers to feature fruit for several different kinds of customers, whether by special labels, brand promotions, or some other approach.[9] Educated consumers can then help translate market requirements into effective, less onerous, cultivation practices.

Other Certified Bananas?

If Fair Trade status is reserved for "disadvantaged producers," what about the corporate role? TNCs have been a major barrier to fairer trade in the past. TNCs can be "traders" or "licensees," as Dole has been in Peru, Ghana, Germany, France, and elsewhere, and as Chiquita is in the Netherlands, Finland, and the United Kingdom. Nevertheless, in principle no transnational firm may be registered as a Fair Trade grower. "Fair Trade and Chiquita respect each other," noted Chiquita Vice President Enrique Vázquez, "but we also know there is a boundary we do not cross" (interview 2004). Yet as consumer interest in fairer trade increases, the alliance must address the expanding market. If the alliance is strong enough to modify corporate behavior to positively improve worker and smallholder livelihoods in reaching this market, it must acknowledge the change. If vigilant

NGOs retain control, FairTrade could involve more extensive corporate distribution of bananas, just as it has with coffee. Laure Waridell pointed out that once consumers made certain that coffee "wears the fair-trade certification logo[,] [w]e can decide whether to buy it from a small company or a bigger one. The more people buy fair-trade coffee, the sooner it will become a must for *all* businesses, large and small" (Waridell 2002, 106, emphasis added). In May 2007, Chiquita sold and leased back its "Great White Fleet," from Eastwind Maritime (New York) and NYKLauritzenCool (Stockholm). The new shipping company could conceivably handle FT fruit as well.

Thus, in open-market conditions, a coherent alliance must relate to banana-certification efforts that stress social and ecological quality. Keeping in mind the five requirements for a useful code noted in chapter 11, we may discover parallel programs that genuinely enhance worker and environmental condition, ones that bolster the attention consumers devote to selecting bananas that are healthy for all stakeholders. How the FairTrade/union alliance credits these efforts will require dialogue with other alternative traders within IFAT and EUROBAN as well as with Social Accountability International, Rainforest Alliance, Ethical Trade Initiative, IFOAM organic grower associations, and smallholders and union producers not covered by any Fair Trade approach.

Such open thinking "outside the box" is not imposed from the North, nor is a capitulation to the neoliberal market. Although dangers cannot be dismissed, such thinking need not surrender Fair Trade aspirations. Given the history of corporate shenanigans, alliance promoters know cross-communication can happen only with transparent and delicate arrangements among all parties. The overall goal must be sustainable conditions, as determined by banana workers and smallholders. As FT advisor Jonathan Rosenthal says, all models of change need debate.

> Right now, I think we have created false dichotomies: small farmers vs. larger farmers, TNCs vs. alternate trade. This is an oversimplification. People have hardened positions with a lot of assumptions. Yet these groups do have the same interests at heart. We must strategize on how to get them together. We need to say to [small Andean producers] El Guabo and Prieto, "You take the lead in designing the model, so small farmers have enough control and power to take risks." So it is not that they trust

us in doing this, it is that we create a safe space to help it occur. Taking such an approach, those involved are already creating relationships: COLSIBA is attending FLO meetings and FLO is involved with EUROBAN. While the exchange is insufficient, it helps build toward common ground. (interview 2007)

These steps would include structural changes in trade agreements and methods of banana production. Yet the outcome could "bear fruit" as Fair Trade/union bananas become a new vision for change.

Summary

I conclude that a FairTrade/COLSIBA/farmer alliance that meets the six challenges of price, union monitoring, consumer education, retailer pressure, smallholder technology, and Fair Trade coordination is feasible. To succeed, it must replicate and publicize the advantages of union-monitored, FLO-approved plantations such Gariba Musah's Volta River Estates in Ghana. As low-cost bananas continue to be shipped from Ecuador and elsewhere, the network must not allow "certified" produce to undercut unions but rather must promote bananas that encourage unionization. Activist consumers can tell Chiquita, Dole, and others to make the right choice in deciding whether to ship from unionized or from unorganized fincas in Guatemala or Cameroon. With effective coordination and educated retailers, the bulk of all bananas exported to the United States and Europe could arrive from Honduras, Costa Rica, Panama, Colombia, and Nicaragua along with an appropriately designated share from FLO-certified Latin, Caribbean, and African smallholders. In addition to the annual verification, unions and producer associations on the ground could offer daily appraisals.

Workers and smallholders could make the crucial difference about these choices if they network with a similar identity and address banana structures. Fair-trade groups; NGO allies like Banana Link, BanaFair, Peuples Solidaire, CTM, Oké, and Oxfam; unions like the IUF, COLSIBA, 3F in Denmark, the CGT in France, and the UFCW, along with the Solidarity Center; smallholder associations like WINFA and CLAC; labor and banana support networks like Foro Emaús, EUROBAN, USLEAP, STITCH, Global Exchange and USAS; and working consumers across the globe can make the coordination happen. They already have in Costa Rica, with the Chiquita and Dole campaigns, and by promoting alternatives to a tariff-only system. With farmers and workers in the lead, even enlightened corporate and government leaders can play a role. Fair bananas have become an invitation for supporters to unite structures, networks, and identity and jointly respond.

Notes

Introduction: Competing Meanings of Fairness

1. Between 1995 and 2006, the total cultivated area for bananas increased by 20 percent to 4.5 million hectares, producing 70 million tons. Of this, 15.5 million tons was exported, 80 percent from plantations of more than 100 hectares that yielded more than 30 metric tons/ha. This compares to the 5–10 metric ton/ha produced by smallholders that grow on 10 ha or less. The largest export growth came from Colombia, Ecuador, Guatemala, and Philippines (see table 1.1). Of those who buy bananas, 34 percent went to the United States/Canada, 28 percent to Europe, 8 percent to Japan, 7 percent to Russia. At 3 percent, China represented a new growth area (drawn from FAO statistics, summarized in van de Kasteele and van der Stichele 2005, 8–10).

2. In a study of U.K. bananas, producers gained 11.5 percent of the retail price, but workers only 1.5 percent. The international trading company took 31 percent, but the retailer retained 40 percent. A similar study in the Netherlands showed the producer making 7 percent, the TNC 29 percent, and the distributor 41 percent (van de Kasteele and van der Stichele, 2005, 32).

3. The 1970s movement for a New International Economic Order for trade was inspired by writers such as Amin (1977). See Emmanuel and Pearce 1972.

4. See Frundt 2005a; Raynolds, Murray, and Wilkinson 2007; and Conroy 2007 for expanded discussion.

5. The author's field research in Central America began in 1995, employing qualitative methods outlined by Travers (2001) and Neuman (2007, 275–342). For an elaboration of the precise use and integration of these methods, contact the author.

Chapter 2. Corporate Banana Structures

1. I use "transnational," not "multinational corporation" since its production and interests rise beyond particular nations. Technically, Fyffes is Irish; Geest is British-based but registered offshore.

2. Financial and supply difficulties led Keith to join with Boston Fruit, a combination of smaller Caribbean shippers founded in 1885 (Myers 2004, 6).

3. Victor M. Cutter, United Fruit president, in Cutter 1926.

4. Gilbert Joseph, Catherine LeGrand, and Ricardo Salvatore (1998) offer a cultural analysis of United Fruit expansion.

5. According to Thomas P. McCann, a former United Fruit official, by 1952 the company held 3 million acres, most of it fallow (McCann 1976, 39–40).

6. Required by Britain to provide mail and passenger transport between England and the West Indies, Fyffes sought United Fruit's help in obtaining a reliable supply of bananas to fill returning freighters. In 1902 United bought 45 percent of Fyffes, retaining its logo (Striffler and Moberg 2003, 11). Although UF let Fyffes run daily operations, in the 1920s, Jamaican producers accused United of deliberately limiting exports to uphold prices, and the government backed the 7,700 strong Jamaican Producer Coöp's independent marketing. Fearing this would become a model, United forced the coop's privatization in 1936 (Myers 2004, 11–14).

7. In 1948, the British government contracted with Antilles Products to handle Windward Island fruit. Geest acquired Antilles Products four years later and rapidly expanded production to replace output the Windwards lost from weather and disease (Trouillot 1988).

8. Britain restricted imports from non-ACP colonies, imposing quotas and a 20 percent tariff on Latin bananas. Belgium, Denmark, Holland, Ireland, and Luxemburg likewise applied a 20 percent tariff. France, Spain, Portugal, and Italy also restricted non-APC bananas. Germany imported without restriction (see chap. 7).

9. Five years earlier, Jacobo Arbenz became president, promising a continuation of democratic reform (Immerman 1982).

10. Then President Arbenz agreed to issue bonds to the company to purchase about 400,000 acres for twice its original price.

11. The rationale for U.S. intervention according to some scholars was more to prevent potential local challenges to elite landholdings than to protect United Fruit interests (see Gleijeses 1991).

12. According to Myers, United reduced its holdings between 1954 and 1984 from more than 1.5 million acres in six countries to 33,000 acres in three countries, but its yields doubled (2004, 48). United retained some lands clandestinely.

13. Ironically, in the mid 1960s, United had attempted to acquire Del Monte, forcing it to expand into bananas and purchase Miami-based West Indies Fruit as a defensive measure (Burbach and Flynn 1980, 208).

14. See Gallagher and McWhirter 1998: S1–S18. As a result of a follow-up lawsuit, the *Cincinnati Enquirer* paid Chiquita $10 million, allegedly for obtaining information and documents about these activities illegally. However, neither the company nor any court denied their veracity.

15. U.S. Ambassador Myles Frechette could find no "example in Colombia of a parastatal security group that has not ultimately operated with wanton disregard for human rights or been corrupted by local economic interests." U.S. Embassy Bogotá, cable, *Botero Human Rights Letter to A/S Shattuck*, Dec. 9, 1994.

16. U.S. Embassy, Bogotá, cable, "Samper Hosts Governors' Meeting on Crime," October 9, 1996.

17. *United States of America v. Chiquita Brands International, Inc., Defendant,* March 13, 2007; Chiquita plea agreement with Dept. of Justice, March 14, 2007.

18. U.S. Embassy Bogotá, cable, "MoD [Colombian Minister of Defense] Alleged to Have Authorized Illegal Arms Sales to Convivirs and Narcotraffickers," April 9, 1997.

19. An investigation by the Organization of American States found that Chiquita/Banadex transported 3,400 AK47 rifles and 4 million rounds of ammunition procured by the AUC (Brodzinsky 2007). Sixty kilos of cocaine arrived at Rotterdam with Colombian bananas.

20. U.S. Defense Intelligence Agency, Intelligence Information Report, "Colombian Prosecutor Comments on Paramilitaries in Uraba," December 7, 1996; U.S. Defense Intelligence Agency, Intelligence Information Report, "Senior Colombian Army Officer Biding His Time During Remainder of Samper Regime" July 15, 1997.

21. U.S. Embassy Bogotá, cable, "Scandal Over Army Request to Convivir in Antioquia," October 8, 1997; CIA, Intelligence Report, "Colombia: Update on Links Between Military, Paramilitary Forces," December 2, 1997. The official military totaled 135,000.

22. U.S. Embassy Bogotá, cable, "Paramilitary Massacres Leave 21 Dead," November 24, 1997.

23. Incoming CEO Roderick Hills claimed he had advised self-disclosure, and U.S. Justice official Michael Chertoff had agreed that stopping payments would jeopardize employees (Lewis 2007).

24. Salvatore Mancuso, a former Colombian AUC paramilitary commander, said his group also received protection payoffs from local representatives of Dole and Del Monte. Both denied the accusations, Del Monte stating it maintained only a local purchasing office (www.alertnet.org, May 18, 2007; bizjournals.com, May 25, 2007); CBS 60 Minutes 2008.

25. These companies are Fresh Cut Fruits, Inc., Baltimore, 1999; Handi-Pak Foods, Inc., Chicago, 1999; Banana Marketing Belgium, NV, 1999, which holds a long-term marketing agreement with CI Bancol, SA, a major Colombian producer.

26. Between 1996 and 1998 in Guatemala, for example, Del Monte/Bandegua increased its independent fincas from 5 to 9. Chiquita quintupled from 4 to 21; however Dole dropped from 10 to 2 (Banco de Guatemala 1998). In Costa Rica, Del Monte/Bandeco added 20 independent fincas, while Dole cut 24 (both companies retained 20 plantations of their own). Chiquita slightly reduced independents from 17 to 13, and its own properties from 33 to 28 (CORBANA 1999).

27. In 1995, five thousand Ecuadorian farms supplied 30 percent of the world market. Only 3 percent of the farms exceeded 100 ha (Myers 2004, 46).

28. Chiquita/United Fruit began banana production in Ecuador in the 1930s. Dole and Del Monte followed, but by 1965, all TNCs had divested themselves of direct production owing to peasant-worker land takeovers facilitated by government policy (Striffler 2003).

29. See various country wage comparisons, Perillo and Trejos 2000; Myers 2004, 47, citing COLSIBA).

30. By 2001, Chiquita dropped nine independent suppliers in Costa Rica; Del Monte lost ten when it demanded consignment contracts. Independent producers Calinda and Grupo Acón inaugurated direct marketing under vaguely stated fair-trade criteria (Eurofruit

2002). In Colombia, Chiquita sold out to independents in 2004, promising to purchase from them for another eight years (ACILS 2006).

31. In Costa Rica, Chiquita buys from the national Caribana consortium, "so large, it is almost a TNC, but it could go elsewhere with its business" (Vázquez 2004). According to its worker Abel Jarquin González, Caribana's Desarollo Agricola Industrial de Frutales, S.A., has four fincas and employs six hundred workers. It makes workers clean five hectares while paying for only four; it sprays continually when workers are in the fields, and does not provide sufficient protective equipment or washing facilities so clothes contaminate others washed at home. Workers suffer skin lesions. The company opposes unions and "fires workers who attempt to organize [despite] many complaints to the Labor Ministry" (Workers of Sarapaquí 2004).

32. Concentration varies by country. Discounters average 9 percent of the European market but reach 33 percent in Germany! The food market is more fragmented in the United States (van de Kasteele, and van der Stichele 2005).

33. This became a justification for Dole's increasing production in Ecuador, Del Monte's in Brazil and southern Guatemala; and Chiquita's divesting itself of 28,000 ha, 1994–2003 (van de Kasteele, and van der Stichele 2005).

34. Contracts account for 80 percent of Dole's North American retail sales, a third to ten clients. In 2004, Del Monte sold 41 percent to ten clients (14 percent to Wal-Mart) (van de Kasteele, and van der Stichele 2005).

Chapter 3. The Fair Trade Alternative

1. In 1982, when indigenous farmers requested help, Franz VanderHoff, a Dutch liberation-theology priest went to rural Oaxaca. UCIRI formed and began shipping organic coffee to northern alternative-trade outlets.

2. Max Havelaar is a character from an 1860 novel by Multatuli—the pseudonym for Eduard Douwes Dekker, a Dutch official in Indonesia who wrote about the cruel exploitation of natives in growing coffee.

3. Thus, labeling efforts built on previous NGO buying arrangements to assure local sellers fair prices or solidarity (e.g., purchasing Nicaraguan bananas during the 1979–1989 Sandinista Revolution).

4. TransFair Germany and EFTA founded TransFair International. Finland developed Kaupan edistämisyhdistysry, based in Helsinki; Sweden, Föreningen för Rättvisemärkt, based in Stockholm (Waridel and Teitelbaum 1999; Almanza-Alcalde 2005). According to Rosenthal 2007, some competition occurred "between the Havelaar approach (every country doing a culturally grass rooted label) and the TransFair approach (the same label everywhere)."

5. The project intended to (1) improve southern capabilities to take advantage of the socially conscious market niche; (2) change the rules for how markets are constructed and how they apportion benefits; (3) raise northern awareness of global inequalities. Fridell believed only the first has been effective, and that only partially.

6. This point was popularized by Lappé and Collins' *Food First* (1975).

7. BanaFair and Gebana began importing fair bananas in the mid-1980s. Solidaridad/Havelaar took an interest in bananas in the mid-90s.

8. See figure 1.1. FOB prices dropped from 24¢ to 14¢/lb., 1963–2003. U.S. import prices declined —2.4 percent a year, 1985–2001. Global trade prices displayed a -.6 real annual decline while consumers paid the same except during a few national retail price wars (FAOSTAT 2005–7; also see UNCTAD *Commodity Price Bulletin* for pricing trends).

9. Though *all* EC members now had to respect ACP protections, Caribbean nations faced wider competition within ACP.

10. Fifty percent was owned by producer groups from Argentina, Brazil, Burkina Faso, Colombia, Costa Rica, Dominican Republic, Ecuador, Ghana, Morocco, Mexico, Peru, South Africa.

11. One extra cost FLO importers faced was the need to buy import licenses under the European banana regime. In 1999, "newcomers" were allocated only 276 tons. To sell more, they had buy licences from other holders at a hefty premium, inflating the price to consumers and reducing payments to producers (FAO 1999).

12. The December 1989 pastoral letter that stimulated formation of the Foro devoted a major section to environmental health. Foro founders included at least fourteen ecological organizations (e.g., Asociación Pro–Desarrollo y Ecología, ASOTRAMA, United Integral, Trópico Cero, AECO, Fundación Guilombé, Nuestra Tierra, SEC), along with campesino, indigenous, labor, religious, and community groups.

13. The Rainforest Alliance conservation NGO was founded in New York in 1987 by people concerned about ecosystems and wildlife. Several factors inspired Chiquita's arrangement with Rainforest, including Alliance concern about deforestation (interview with Wille 1999), and company fear of militant unionism (interview with Holst 1999; Interview with Wunderlick 1999). RA created the Sustainable Agricultural Network (SAN) in 1991 and certified its first farm in 1993. SAN emphasized three areas: conservation of habitat and the reduction of chemicals; good treatment of the workforce, including minimum wages, good housing, and space to organize (while recognizing Solidarismo, it claimed organizational neutrality); and a management and monitoring farm plan (Divney 2005). See also Chiquita Brands 2000; Taylor and Scharlin, 2004; and chapter 11.

14. Inspired by tea conditions in 1997, FLO created criteria for hired workers and eventually established a Standards and Policy Committee, with representatives elected from participating stakeholders, that reviewed changes triennially. Their suggestions evolved into a list that producers endorsed and FLO certifiers followed.

15. Although we discuss both here, the standards for small farmers' associations separated Standards on Labor Conditions from Social Development Standards (see FLO 2004).

16. In 2005, volatility in the coffee market brought below-market prices to certified coops, causing some to abandon production (Jaffee 2007). For FLO's view on producer impacts, see http://www.fairtrade.net/impact_studies.html.

17. Importer/wholesalers and retailers absorb 60 percent of the banana retail price. Transport costs 12 percent. TNC exporters take as much as 21 percent of the remainder, whereas Fair Trade producers and Fair traders share it (author calculations; van der Kasteele and van der Stichele 2005, 32).

18. Some fair-trade NGOs only partially participate in FLO. In Japan, since 1989 Alter-Trade has imported bananas from the Philippines. CTM Altromercato in Italy and Bana-Fair in Germany also sell some non-FLO bananas verified by their own standards.

19. In 2005 TransFair certified 3.6 thousand tons (more than 7 million pounds). In 2006, it licensed AgroFair's acquisition of a million pounds from Peruvian smallholders and 2 million pounds from Ecuador. It collaborated with Turbana/Uniban/Fyffes (Colombia) to ship 214,000 pounds. Its certifications rebounded somewhat in 2007.

Chapter 4. Actors for Banana Development

1. For meanings of sustainability see Schroyer and Golodik 2006. Denis Goulet adds enhanced self-esteem and freedom to renewable food, shelter, health, and security. Sustainability also encompasses an absence of violence and intimidation so that people possess wider choices to influence their future (Goulet 1989).

2. These categories reflect various nomenclatures: for example, Jan Kippners Black (1999) contrasts harmonic and discordant interests; Barbara Thomas-Slayter (2003) distinguishes neoclassical and political-economic approaches.

3. Modernists usually view government-controlled development as overly bureaucratic and inefficient, whether in socialist societies like the former East Germany or in nationalist-oriented countries like Peron's Argentina.

4. Danaher 1994. Despite exceptions, the trend has persisted. In 2007, the World Bank claimed to rediscover the importance of agriculture.

5. See Cockcroft, Frank, and Johnson 1971; and Wallerstein 1976 on dependency.

6. Rice as quoted by Rosenthal (2007).

7. This attribution of value to price rather than to the labor that produced the good is what Marx called the fetishism of commodities.

8. EUROBAN members: Austrian Banana Campaign (Austria); IFOR Lateinamerika Komitee (Austria); Max Havelaar Belgie (Belgium); Oxfam Wereldwinkels (Belgium); Oxfam Solidarité (Belgium); Banana Link (Britain); Fairtrade Foundation (Britain); Christian Aid (Britain); Oxfam (Britain); U.K. Food Group (Britain); International Centre for Trade Union Rights (Britain); World Development Movement (Britain); IBIS (Denmark); Confédération Générale du Travail (France); Réseau Solidarité/Peuples Solidaires (France), Solagral (France), RONGEAD (France); FLO International (Germany); Bana-Fair (Germany); Die Bananen Kampagne (Germany); Irish Fair Trade Network (Ireland); Banana Watch (Ireland); Centro Nuovo Modello de Sviluppo (Italy); Consorzio CTM Altromercato (Italy), Cooperativa Commercio Alternativo (Italy); Centro Mondialita Sviluppo Reciproco (Italy); FNV (Netherlands); Solidaridad (Netherlands); Max Havelaar Foundation (Netherlands); Stichting Rechte Banaan (Netherlands); Plataforma Rural (Spain); COAG (Spain); CERAI (Spain); Naturskydds Foreningen (Sweden); Paso Global (Switzerland); Gebana (Switzerland); IUF (Switzerland).

9. As predecessor to the Solidarity Center, the unabashedly pro–U.S. government American Institute for Free Labor Development (AIFLD) manipulated local unions for political purposes (see Shorrock 2003; Gacek 2005).

10. It is rooted in the famous international solidarity struggle led by Guatemalan Coca-Cola workers to maintain their union in the face of ruthless repression (Frundt and Chinchilla 1987; Levenson-Estrada 1994).

Chapter 5. Going Bananas as a Social Movement

1. This summary expands on the political process model, which integrates political opportunity structures, mobilizing structures, and framing (McAdam, McCarthy, and Zald 1996; Tarrow 1998; Tilly 1978).

2. These writers examine identity formation and the symbolic expression of shared meanings that often impact coalitional campaigns (see Touraine 1981, 1985).

3. These points draw on Conroy 2006.

4. This is a TransFair assessment (Raynolds 2007). Caribbean nations could possibly supply bananas if appropriate pricing and shipping mechanisms were created.

5. While COLSIBA is not an official IUF member, half of its unions are.

6. Here networks are considered within the social-movement theory of resource mobilization as essential resources, but there are other resources (Zald and McCarthy 1979).

7. For network applications, see Dreiling and Robinson 1998; Ayres 2002.

8. Nevertheless, transnational activists can encounter the difficulties summarized by Brandy and Smith (2005, 231–52), such as achieving equality between first and third world alliances (Wood 2004). Sikkink (2004) acknowledged that complex interactionism is required.

9. Such networking occurred with IUF responses to Coca-Cola's human rights violations in Guatemala (Frundt 1987, 2003), or negatively, with AIFLD manipulations of Latin unions.

10. Polletta (1999) has refined the concept of free space as a network intersection critical to generating mobilizing identities.

11. The framing process that Goffman and others delineate is used for more than just identity formation.

12. Borrowing from cultural theory, identity theorists analyze organizational language for insight into how identity can be manipulated. Movement, corporate, and government discourse validates certain actions more than others. In analyzing discourse "apparently designed to ameliorate environmental destruction," Brosius (1999) discovered successful efforts to "obstruct meaningful change through endless negotiation, legalistic evasion, and compromise among stakeholders." The result imposed a type of identity as it naturalized "a discourse that excludes moral or political imperatives in favor of indifferent bureaucratic and technoscientific" intervention.

13. In the process, they may discover that the status of an affiliate's other identities may be complex or confusing. A forthright response to the question, "What does it mean for you to be a member of X farmer association?" or "In what way do you see yourself as a member of Y union?" would likely depend on the respondents' particular experiences, not on their appraisal of the general benefits of collective action. However, additional questions could test their ethnic interaction, environmental consciousness, gender acknowledgement, and openness to international organizations.

14. COLSIBA questioned certification of any producer or coops that hired contract labor, a point pursued below.

15. Assessing coalitional identity over time often means clarifying its connection to movement networks and opportunities. As a social-movement network grows, identity may be expected to expand unless separate interests remain clearly complementary. As a movement gains momentum, common identities would usually broaden, assuming movement leaders perceived resource opportunities that supported the expansion.

Chapter 6. The Persistent Banana Environment

1. For years bananas were shipped as stems. Dole introduced boxes in the 1960s.

2. Companies later explored genetic modification, such as Rahan Israeli technology to prevent nematodes and nematicide applications.

3. This is an acronym for 1,2-dibromo-3-chloropropane (CAS #96-12-8), a highly toxic nematicide.

4. Chiquita stopped use of the fungicide Bravo (chlorothalonil) by 1990, but others still spray it.

5. The nematicides cadusafos, carbofuran, and ethoprophos severely harm fish. Chlorpyrifos, tiabendazol, propiconazole, and imazalil also frequently haunt Honduran and Costa Rican costal waters, and ditches in Ecuador (Harari 2005a, 45–46).

6. Munguía demands that "banana importing countries . . . reject the indiscriminate use of chemicals that affect women's reproductive health," and quickly resolve DBCP "lawsuits relating to sterility or other diseases" (2005, 88–89).

7. In Costa Rica, Foro Emaús aided workers sterilized by DBCP, eventually filing for damages in U.S. courts (see Barrantes Cascante 1999, and helpful elaboration in Sass 2000).

8. It aerially sprayed propinconazole and benomyl (an EPA-listed potential carcinogenic); mancozeb, azoxystrobin, thiophanate-methyl, tridemorph (dangerous for fish), and bitertanol (banned in the United States).

9. See Ministerio de Salud, Costa Rica 1998, tables 11, 15, 18, 23–24.

10. ISO 14001 was designed by private-sector representatives from developed nations to voluntarily improve environmental-management systems.

11. The TNCs restricted the Pesticide Action Network's dirty dozen and the World Health Organization's acutely toxic list, yet often disregarded cumulative chemical effects (Harari 2005a, 47).

12. Union contract enforcement helped field hands in Costa Rica, Honduras, and Panama avoid direct spraying, but not those in Guatemala and Ecuador.

13. The Escuela de Agricultura de la Region Tropical Humeda estimated that in each Costa Rican production cycle fungicides were aerially applied forty times. Forty percent hit the soil; 15 percent was lost to wind; 35 percent reached the plants but was washed off by rain, totaling a 90 percent loss, in this case of 11 million liters of emulsion (cited by Chambron 2000; see also Hernandez and Witter 1996).

14. In 2000 Rainforest responded to criticisms of its social verification by emphasizing that its approach was primarily environmental (Chiquita had also adopted the SA8000 social audit, chap. 11.) However, Rainforest reasserted social claims in 2006.

15. Chiquita's corporate responsibility officer David McLaughlin estimated an 80 percent reduction in herbicides, a 50 percent reduction in nematode control, and a 50 percent drop in postharvest fungicides that saved the company over $1 million annually. Only 10–15 percent of chemicals subsequently used were EPA class 1 or 2 [highly toxic], and plant-root health increased (Taylor and Scharlin 2004, 102, 75).

16. Verified by McCracken (interview 2004), who found individuals deciding not to use equipment because it was poorly designed, or uncomfortable in hot conditions.

17. They include such points as: complying with national legislation on worker security and health; clearly defining worker responsibilities; guaranteeing adequate training and risk vigilance; continual improvements to reduce labor risks; revisiting objectives to assure implementation; and promoting this policy at all levels to workers, providers, and contractors.

18. Flooding swept numerous bags toward populated areas in Sixaola, Costa Rica, in January 2005.

19. In 2007 in the Philippines, companies resisted a Davao city ordinance restricting aerial spraying.

20. Hermocilla (2004) referred to media reports of a 1.4-meter mound of dead fish around Matina Costa Rica due to nematicide spraying. Industry claimed a purposeful discrediting of Costa Rican bananas—just as European sales were increasing (articles by Marvin Barquero and ads in *La Nación*, San José, May 2004). The incidents were coincident with the rains, which came just after spraying had occurred (interview with Hermosillo 2004).

21. *Freshplaza*, September 6, 2006. *Freshplaza*, http://www.freshplaza.com, is a global independent Internet news source on the fruit and vegetable sector. It appears in various languages and is edited from the Netherlands.

22. These include any of the Pesticide Action Network's "dirty dozen" and those rated class 1a and 1b by the World Health Organization.

23. For example, Thiobenzadol must be mixed with 50 percent less persistent agents to keep residues 25 percent below regional non–Fair Trade levels. FLO criteria also removed chlorine from postharvest baths, and banned painting stem crowns with Tremux-like chemicals that cause rashes on packinghouse women.

24. Partly for its climactic advantages, the Dominican Republic recently became the world's leading exporter of organic bananas.

25. FLO insists that no plantings occur on inclines of greater than 60 degrees, and that areas between 30 and 60 degrees be contoured to reduce runoff, with drains dug along the contours, and grass or trees planted every 7.5 meters or so between the drains. Certifiers make sure that channels keep water a meter below land levels. Without using herbicides, this requires a fair amount of hand labor.

26. To protect the waterways, FLO producers plant trees and vegetation in twenty-meter zones along rivers (narrower if no aerial spraying occurs, but potable water supplies

require a hundred-meter buffer). Two-meter strips along channels catch and filter field runoff and help reduce chemical contamination.

27. Smith (1994) describes a Honduran effort to recover ancestral farm traditions.

Chapter 7. Conundrums of the Banana Trade

Author's note: Portions of this chapter appeared in Frundt 2005a.

1. Lomé was renewed in 1980 and 1985 for five years, and in 1990 for ten years.

2. EU nations had been purchasing 40 percent of the world's bananas via two systems (chap. 2): Britain, France, and Spain paying higher prices to current and former (ACP) colonies; others, like Germany, purchasing on the open market. When Europe moved toward a unified market in 1993, Regulation 404 imposed a complicated general arrangement of licenses and quotas.

3. The original allocation of 2 million tons with a 100 Euro/ton tariff was soon changed. When the EU expanded in 1995, Latin bananas received a 2.5 million-ton quota. Excess fruit was effectively excluded by a $680–$822/ton tariff.

4. Individual ACP nations could count on an export quota assured by Lomé Article XIII. Thirty percent of the Latin quota was licensed to ACP handlers: 57 percent for primary country importers; 28 percent for ripeners/distributors; 15 percent for secondary importers. The Union offered licenses at no charge but gave preference to European applicants.

5. Chiquita claimed losses due to its higher tariff costs in former free-trade nations like Germany, and to the licensing system, which gave 30 percent of its usual Latin shipments to ACP importers; however, a 1995 FAO study found that Chiquita's EU market share between 1991 and 1994 only fell from 25 percent to 18.5 percent (Myers 2004, 76).

6. Although Dole and Del Monte did not suffer from the EU's new tariff-quota system, they officially indicated agreement with Chiquita's action (Taylor and Scharlin 2004, 139f).

7. Also, TNCs now had to pay the 75 Euro/metric ton tariff on Latin quota imports to *all* European countries.

8. They also argued that Regulation 404 encouraged large European firms to quickly set up "efficient" plantations (more productivity per acre/hectare) in ACP countries to compete with Latin American bananas, thereby undermining smallholder production. Yet TNCs began doing the same thing (Astorga 1998; van de Kastelle 1998).

9. M. Garcia, head of Del Monte's Guatemalan operations (2005).

10. An increased share would also earn a higher base for license allocation.

11. It sought a GATT waiver for economic and social considerations under Article XXV.5 requiring a 75 percent membership approval, which it received by consensus. The United States later contested the waiver under WTO procedures.

12. TNCs opposed the framework clause that partitioned Latin nations, giving Costa Rica and Colombia more than 45 percent of the Latin allocation and excluding Panama, a major Chiquita supplier. The partition affected workers because it "caused a new phase of competition, stimulating U.K. transnationals such as Geest and Fyffes to enter these

Latin countries and imitate the anti-union actions of Chiquita, Dole, Del Monte [by] . . . eliminating many of the social guarantees for workers. Companies refused to negotiate contracts, reduced salaries, increased the working day, agitated for union rejection, stepped up union persecution, and abandoned fincas without paying worker obligations" (interview with Bermúdez 1999).

13. In addition, Chiquita CEO Carl Linder was a major contributor to both political parties, second among Republicans in 1999–2000 (Bartlett and Steele 2000, 42f). Even in the context of other agricultural lobbying efforts, the journalists judged the government's intervention on bananas to be notably unusual.

14. The panel concluded that the distribution of Latin import licenses to French and British firms took away a major part of historically U.S. business. The panel did say that GATT Article I allowed country-specific quotas to substantial suppliers as long as it recognized all substantial suppliers [e.g., Panama, Ecuador]; and that GATT Article XIII ACP quotas were legitimate.

15. Under U.S. influence, the appellate body went beyond the panel and removed Article XIII country quotas. Disregarding the harm to smallholders, the body disbarred Caribbean representation (Myers 2004, 89–90).

16. President Clinton himself chimed in: "The Europeans are basically saying, 'You've won this trade fight under the law, but we still don't think you have a meritorious position; therefore, we will not yield.' Well, when we lose trade fights, we lose them" (Clinton 1999).

17. Without waiting for WTO approval, the United States began requiring penalty-equivalent bonds on products from countries favoring the regime: cashmere from the United Kingdom, French handbags and bath preparations, Italian cheese, and electric coffee makers, paper goods, sheets, and industrial batteries.

18. The panel compared Chiquita's income to what an alternative regime would have generated. Quotas could be allowed if all parties could agree on a reference period for basing trade calculations. Ecuador was the first developing country given the right to impose sanctions, but it never did so, fearing greater punishment to its economy (Myers 2004).

19. Besides having the ear of USTR and Sen. Robert Dole, Chiquita gained specific legislative authority via the Africa Growth and Opportunity Act to affect how sanctions were rotated among countries.

20. One Netherlands study estimated that under the First Come, First Served system, shipments from Costa Rica would drop from 25 to 11 percent of the EU market; from Panama, from 16 to 8.5 percent; from Honduras, from 9.4 to less than 1 percent. Colombia would lose 25 percent of its export sales. However, Ecuador's share of the market would increase from 26 to 62 percent (cited in USLEAP 2001a).

21. Whereas Del Monte sourced 13 percent there, and Chiquita, 7 percent (Frank 2002, 1).

22. Although supported by the United States, the waiver was not granted until November 2001, with the understanding that any new tariff level "would result in at least maintaining total market access" for all WTO member suppliers. Appended arbitration rules would become very important by 2005 (Myers 2004, 122).

23. Fair Trade Labeling Organization International (FLO) to College of EU Commissioners, Apr. 23, 2001. At that point, FLO, a member of EUROBAN, represented twenty-one organizations in seventeen countries that shipped 25,000 tons of bananas.

24. European supermarkets inaugurated the Eurepgap standard. In March 2004, the U.S. Food and Drug Administration required importer registration of fresh produce, with similar effect (chap. 11).

25. By 2004, Latin exports to Europe included 211,000 tons from Costa Rica, 173,000 from Colombia, and 56,000 tons from Brazil, compared with 150 tons from Cameroon (Garcia 2005).

26. The listed market price is usually "Free on Board," (FOB) which includes internal transport costs, palleting, and loading boxes on ship. When importers deal with exporters rather than directly with producers, "farm gate" is the relevant fair-trade price (what is paid to Finca 6). Whether FOB or farm gate, occasionally, the price does fall below the market price, as happened in Ecuador in early 2006, when import contracts held the price at a lower level. After a stinging lesson about timely adjusting coffee payments for inflation, FLO quickly worked to correct this.

Chapter 8. Resilience of Banana Unions

1. COLSIBA referred to Fair Trade in its 1992 founding document.

2. Yet during negotiations on ILO Convention 184, employers cited requirements for proper sanitary facilities as "outrageous" (Longley 2005).

3. On AIFLD, see Spalding 1977; Armstrong, Frundt, Spalding and Sweeney 1987; Frank 2005.

4. "Fordist" refers to the extended package of worker benefits that Henry Ford offered his auto workers in exchange for company loyalty and assembly-line work. In the Latin banana sector it also meant that if a company provided work, the state agreed to incentives (Harari 2005b). The U.S.–sponsored coup in 1954 rejected Guatemala's Fordist reforms.

5. In the early 1990s, thirteen Guatemalan contractors formed COBIGUA, which produced exclusively for Chiquita. UNSITRAGUA sought to organize COBIGUA's fincas.

6. The company claimed a $10 million loss, and estimated another $18 million for rehabilitation (Taylor and Scharlin 2004, 167), influencing its decision to sell to the workers.

7. Chomsky (2004b) cites extensive Colombian documents and direct participant testimonies to support this.

8. To avoid trade penalties in 1993 (and further amended in 2000), Costa Rica's Solidarista Act restricted the associations to social functions, but it still allowed "permanent committees" of three workers to sign agreements unless a union garnered 34 percent of the workforce, sufficient to negotiate an enforceable contract. A campaign in Britain and Costa Rica by the World Development Movement, Banana Link, and the SITRAP union persuaded Del Monte–CR to sign a "Framework Agreement" in December 1997 that allowed free organization. "SITRAP members had the majority in one third of the elected permanent committees on Del Monte's 24 plantations, and . . . rose to over 20 percent of Del Monte's total workforce of 4300" (Banana Link 2003).

9. Del Monte had not expected "so many workers to freely choose the trade union" instead of Solidarista, so it fired SITRAP activists (Banana Link 2003).

10. U.S. Embassy, Bogotá, cable, 1996. "40,000 Colombians March to Protest Wave of Kidnappings; Paramilitary Group Releases Two Guerrilla Relatives," December 2.

11. Interview with Bermudez 1999; and author's personal observation.

12. COLSIBA, point no. 5, Statement of Principles, supports bananas produced in a just and environmentally sensitive way.

13. The union of semi/nonspecialized workers of Denmark.

14. Workers had originally demanded job restoration, improved wages, health benefits, housing, and schools. After the company restored worker jobs, the Honduran government declared the strike illegal; but smarting from earlier layoffs, union members voted to continue with additional demands. COLSIBA aided the settlement.

15. Author interview with Honduran workers, 1999. Banana Link's *Trade News Bulletin*, no. 21 (1999) argued that monocultural production worsened the hurricane's impact.

16. Del Monte hesitated several days before rejecting SITRABI's forced resignations (Perillo 2005).

17. The workers did have to accept a 70 percent cut in health benefits, 30 percent wage reduction, and losses in education and housing (USLEAP 2000).

18. Author interview with SITRABI Executive Committee, June 10, 2002 (Noe Ramirez, Ernesto Soral, Edgar Agusto, secretary of conflicts; Jesus Martínez Sosa, secretary; Selfa Sandoval, secretary of organization).

19. Founded in 1919 by representatives from governments, companies and unions, the ILO became a U.N. body in the 1940s. Over the century, the organization assessed work conditions around the world. Delegates from its many participating nations (50% government, 25% union, 25% business) endorsed a series of fundamental conventions and recommendations. These can be accessed at http://www.ilo.org/global/What_we_do/InternationalLabourStandards/index.htm

20. Verifying such an "understanding" has proved difficult for some FLO certifiers (see chap. 12).

21. The proposal is detailed in COSIBA-CR 1999, signed by the five CR union leaders.

Chapter 9. Peasants of the Caribbean and Fairer Trade

1. St. Vincent average yield is less than 7 tons/acre, but some reach 12 tons (Rose 1994).

2. Ironically, while a stable Caribbean banana supply has been an issue, its recent production of Fair Trade bananas often exceeds demand, as I discuss below.

3. According to Bowman, it takes six to seven years to obtain organic certification. Farmers need intermediate assistance handling predators and production losses (interview 2005).

4. Despite investment and output reductions in the 1990s, unionized Jamaican plantations remained active. According to Clifton Grant, second vice president of the

IUF-affiliated Allied Workers Union, thirteen hundred of its workers on two large plantations handled 90 percent of exports (2005).

5. See Myers 2004, 25ff for this history.

6. Dominica forced Geest to sell land to farmers directly (Crichlow 2003).

7. In fairness, Geest was complying with *Codex Alimentarius* and European Union requirements that first-class bananas be a minimum of 14 cm long, etc. U.K. supermarkets had even more rigorous standards.

8. For Senator Josephine Dublin-Prince, "the banana is synonymous with both economic and social development.... [With] 51 percent of the population ... female and 38 percent of households headed by women," their "caring and nurturing" is critical. "60 percent of field tasks essential to the viability of the industry are done by women, yet they collectively receive less than half of the wages" (2005, 90).

9. EU Reg. 404/93 also induced companies to diversify ACP sources to gain additional import licenses, increasing investments in the Dominican Republic, an ACP State as of 1990. TNCs further exploited ACP status by establishing plantations in ACP countries lacking labor protections, such as Cameroon and the Ivory Coast.

10. Besides weather, the EU policy changes effected this decline.

11. For example, in St. Lucia the TQF, tied to the United Workers Party, offered better prices to more "efficient" producers, wooing them from the small-farmer SLBC, linked to the Labor Party.

12. According to Grossman's careful counting in the Restin Hill area of St. Vincent in 1988–89, 97 percent of households had used at least one pesticide. For bananas, the herbicide of choice was paraquat (1998, 197).

13. Supermarket demands for standardized bananas inspired the Certified Growers' Program (chap. 11).

14. Cargill Technical Services, *Socio-Economic Impact of Banana Restructuring in St. Lucia* (Minneapolis: Cargill Corporation, 1998), cited by Liddell (2000).

15. Windward governments backed technological enhancements that risked transforming smallholder production into miniature monocropping enterprises. St. Lucia's Ministry of Agriculture advocated *meristem cultivars* (a plant variety of undifferentiated, more adaptable cells) that necessitate pure stands and more chemicals. The St. Vincent government emphasized high-yield varieties and irrigation.

16. Surrendering its role as a Caribbean banana-shipping agent, Geest expanded direct production in Costa Rica, but it also lost revenues there.

17. TNCs remained skeptical of this: While "60 percent of Swiss consumers *are* willing to pay"; Germans "say they will buy," but "60 percent patronize discounters" (Cuperlier 2005).

18. Moberg discovered retail U.K. FLO bananas to be priced 119 percent higher than generic bananas, with growers gaining only 41 percent of the differential (2005, 10).

19. WINFA averages combined production expenses from each island. FLO checks market prices with national importers to determine if the average is reasonable, but this can take several months. Dissatisfied producer groups may file a complaint with FLO's Standards and Price Committee, which can adjust the price but it takes time. Price is also

affected by FLO's own fee, based on audit costs. An association that consistently meets FLO's criteria needs fewer audits and gets charged less. FLO acknowledged its pricing system is not optimal. In 2005, after losing coffee affiliates owing to its policies, FLO inaugurated a more precise and timely appraisal process.

20. In 1995 marijuana overtook bananas as a primary contributor to the St. Vincent economy, involving 12,350 acres and a notable number of former workers and growers (see James Anderson, Associated Press Wire, December 4, 1998, published in *Marijuana News*). Subsequent eradication efforts by the U.S. military have not altered its status, given the region's high unemployment.

21. Organic trade expanded 15 percent to reach 1 percent of world consumption in 2003; 55 percent of the organic trade went to Europe (van de Kasteele and van der Stichele 2005).

22. FAO 2003 compared prices consumers paid for organic Fair Trade with conventional bananas, and then calculated the percentage increase received by growers. In some nations, the high price differential was not proportionately shared with small growers.

23. IFOAM (International Federation of Organic Agricultural Movements), constituted by 720 organizations in ninety-eight countries, develops, revises, and harmonizes organic standards and criteria, and lobbies governments to conform their organic regulations to the standards. IFOAM Sec. 8 guidelines include recommendations on wages and social rights, but these are not mandatory.

24. Half of the thirty-nine producers randomly sampled understood what Fair Trade was (Shreck 2002).

25. Another FLO-certified Association of Ecological Bananas of the Northwest Line (Banelino), an affiliate of AgroFair's co-op network CPAF north of Ázua, claims some identity success in uniting its three hundred small producers. As of 2008, http://www.FLO-CERT.org (operators list) displayed fifteen additional certified Dominican producers.

Chapter 10. The Chiquita Accord and Labor Responses

1. "Agreement on Freedom of Association, Minimum Labor Standards and Employment in Latin American Banana Operations," signed by Steven Warshaw, chief operating officer, Chiquita; Ronald Oswald, general secretary, IUF; German Zepeda, coordinator, COLSIBA; with ILO Director Juan Somavía as witness, June 14, 2001.

2. Chiquita declared bankruptcy in March 2002, reorganized, and began afresh in May 2002.

3. Taylor and Scharlin 2004, 124.

4. "In Sarapiquí, Costa Rica, we have had four or five meetings, one just yesterday that lasted 9:00 a.m. to 8:00 p.m. We discussed nonfunctioning equipment, fumigation over worker areas, housing issues, and improved salaries for the workers. Discourse does gain positive things, but issues are often not resolved" (interview with Barrantes 2004).

5. One hectare equals 2.2 acres. Pay may be increased to 1,700 colones when infestation is high.

6. The following is based on author interviews of workers and leaders in Costa Rica and Guatemala, 1999–2004; and on Harari's survey of banana unions (2005b).

7. According to Vice President Vázquez, "Chiquita [like the others] faces two problems: first, the bigger producers like Caribana in Costa Rica are in a position to set their own terms when we need their fruit; second, the smaller producers are often stretched to near bankruptcy, which they take out on the workers. In the second case, we give the owners a lot of support, and they owe us a lot. But some are not going to make it, and the workers will lose their livelihoods" (interview with Vázquez 2004). The unions suspect a gentlemen's agreement to preserve exploitation.

8. In late 2000, Chiquita stopped buying bananas from fourteen independent producers and announced it would not recontract six in 2001. Dole cut purchases from fifteen independent producers by 25 percent and did not renew six contracts in 2001. Del Monte refused to recontract with any independent producers in 2001 (Taylor and Scharlin 2004, 167–68).

9. "Ironically some like the three-month contract because it gives them a fifteen-day break for going to the doctor, or being with their kids. Visiting the doctor during the regular work period goes on their record, and too many absences mean they will not be recontracted" (interview with McCracken 2004).

10. Chiquita "receives monthly reports through our human relations office that document this reduction in rotation" (interview with Vázquez 2004).

11. Harari 2005b, survey tables. Honduras retained a multitask approach.

12. According to *Banana Link*, the Chiquita agreement brought concrete improvements in Colombia and Honduras but worked effectively only where unions were still strong enough to negotiate. Its holistic approach covered all plantations but implementation among suppliers remained weak (Chambron 2005).

13. Chiquita also sought to replace some traditional paper box packaging with reusable plastic containers, generating protests from SITRAIBANA and SITRABI—the union of workers for Chiquita-Panama's independent producers (*Freshplaza*, September 12, 13, and 21, 2006).

14. In Bocas, COOBANA lost $135,000. In Armulles, SITRACHILCO filed complaints before Panama's Competition and Consumer Affairs Commission and Human Rights Ombudsman that the agreement had endorsed "corrupt contracts" in establishing COOPSEMUPAR as a cooperative in 2003 (*Central American Report*, January 27, 2007, 3).

15. Following a special mission convened by COLSIBA and joined by supportive NGOs including USLEAP, Chiquita signed a refresher agreement with SITRATERCO in Honduras (USLEAP 2006b).

16. Harari (2005b) claimed that, with the exception of Nicaragua, maternity and reproductive health rights are generally not acknowledged.

17. EUROBAN letter to Chiquita, September 2006.

18. In addition to previous SITRABI efforts, Danish and Belgian unions began organizing projects on Guatemala's southern coast in 2007.

Chapter 11. The New Banana-Marketing Strategies

1. Northern-controlled ISO applications are expanding (ISO 22000 for food safety, ISO 28000 for bioterrorism tracing, ISO 65000 for health systems), and FLO-CERT has

cooperated. For trust and equity reasons, however, FLO has resisted applying ISO 26000 to standardize social-responsibility accounting (Buitrago 2007).

2. By March 2004, all U.S. produce importers had to register with the FDA via a complicated computerized procedure. Small producers found it necessary to apply jointly to obtain a traceable registration number for each shipment.

3. In the United States, *Forbes* magazine gave Whole Foods high ratings for employee relations. Yet, since 2000, it has militantly resisted a United Food and Commercial Worker organizing drive.

4. Although Chiquita claimed to "heed a warning from BanaFair Germany that we should not cross into their own territory" (interview with Vázquez 2004), it signed with TransFair USA in 2005. Dole began shipping FLO-approved bananas in 2003.

5. Fair Trade spokespersons question CORBANA's assessment of this interaction.

6. MIT Professor Dana O'Rourke showed how Price Waterhouse monitors in China accepted management's assessment regarding freedom of association (Independent University Initiative 2000). Other groups have challenged positive monitoring reports on Nike in Indonesia that bypassed persistent verbal and physical abuse (Yanz and Jeffcott 2001; Esbenshade 2004). International nonprofit monitors may treat labor issues more objectively, despite difficulties all monitors face in China (MSN 2001).

7. Workers Rights Consortium and Fair Labor Association auditors have offered mechanisms for public transparency in the maquila sector.

Chapter 12. Fair Trade and Freedom of Association

1. For example, in 1824 workers forced Britain to repeal the Combination Act of 1799.

2. Although the United States has not ratified ILO Conventions 87 and 98, it has incorporated these principles into its own labor legislation.

3. When critics argued that killing trade unionists prevented their ability to organize, USTR officials retorted that such killings had to occur while the victims were directly engaged in legal union activity, *and* the perpetrators had to be shown to retaliate for that reason (Frundt 1998)!

4. Leaders may be legally protected, but it requires extensive litigation.

5. See Frundt 2004a and Seidman 2007 for FOA evaluations in practice.

6. "FLO has cooperated with ISEAL, a program on Social Accountability in Sustainable Agriculture. ISEAL is a collaboration of international standard setting and accreditation organizations" that considers smallholder access to certification and opportunities for convergence (Chambron 2005).

7. Fair Trade bananas constitute an important market segment in a number of EU countries, as much as 50 percent in Switzerland, and so on (see table 3.2). Yet some Fair Trade certified producers pay little attention to labor rights and "buy bananas from cooperatives who pay other workers below minimum and violate rights" (interview with Bermúdez 2004). For a synopsis of some TransFair-approved producers, list the country on the Google Earth Web page, then layer "global awareness," and double-click to choose "Fair Trade Certified."

8. Shreck 2002 and follow-up phone interview with author, 2007.

9. Between 2000 and 2003, organic yields in Costa Rica achieved a 36 percent increase. Bananas cost 8.7 percent more to produce but gained a 42¢/box premium (FAO 2004).

10. Unions also charged non-FLO Costa Rican producers with freedom-of-association violations, including the Earth University/Whole Foods "Earth First" brand producers and Caribana. Hermosilla (interview 2004) cited the Voelker plantations near Panama, which "reduced chemical applications for the European organic market, but immediately fired workers that tried to organize." McCracken (interview 2004) described "Acon's fourteen fincas near Limón that shipped 'Tropical Best' to Belgium stamped 'Costa Rican Bananas are environment and worker friendly.' Acon has no certification."

11. Producers must cover costs for the fertilizers, bags, boxes, liners, stickers, fungicide treatment, and waste disposal and deliver four fingers per hand in each box.

12. As a "trader," Dole does not have an obligation to recognize a union, but following FLO criteria, it is to respect ILO Conventions 89 and 98.

13. A Finnish TV crew exposed the owner's low wage payments and misuse of the premium on a farm that produced 40 percent of Finland's FLO-certified bananas (Taylor and Scharlin 2004, 194).

14. The Association of Small Organic Banana Producers of Saman (Asociación de Pequeños Productores de Bananas Orgánicos de Saman y Anexos). (See http://www.appbosa.com.pe.)

15. With 1.4 ha. each, APPBOSA's 153 members shipped 6330 metric tons in 2006, doubling the output of APOQ's 317 members in Querecotillo (FLO 2006a, 2006b). (See also notes 7 and 14.)

16. Also includes APROVOPCHIRA.

17. In its 2006 general assembly in the Dominican Republic, CLAC reiterated that the function of TNCs "goes against the principles of Fair Trade," which represent "a different type of marketing." CLAC warned about TNC "capacity to take advantage of the new sensibilities of consumers," adding that "it does not serve to simply stretch the market thoughtlessly, but rather [it is necessary] to really help the small producers in their efforts to survive and develop" (CLAC 2006b, Resolution #8). Nevertheless, as Lucio (2007) indicated, some adaptations may be in order.

18. Some see a proliferation of Fair Trade organizations; others, a consolidation. As I indicated in chapter 3, early religious-based, fair-trade groups created market outlets. Nationally organized nonprofits such as CTM and the various Oxfams operated with some autonomy. FLO has aided many such groups to come to agreement on common standards.

19. For example, the Fair Trade Foundation expressed "shock and dismay" about conditions European migrants faced at Pratt's bananas, where they worked in long shifts without bathroom breaks. Banana Link attributed this kind of exploitation to supermarkets squeezing their supply chains (http://bbc.co.uk, May 2007). FTF warned about their discount banana pricing (2008).

NOTES TO PAGES 189–196   241

Chapter 13. Dole: Reluctant Fairness

1. In 2006 the Philippine Amado Kadena–National Federation of Labor Unions–Kilusang Mayu Uno (AK–NAFLU–KMU) struck Dole Philippines, Inc. (Dolefil) in a wage-and-benefits dispute. Claiming sabotage, Dole filed notice of a lockout. The strike lasted five months until the Philippine Department of Labor and Employment assumed jurisdiction after official mediation failed (*Freshplaza* [Netherlands], September 25, 2006).

2. Although on the SAI board, Dole did nothing when a sourcing producer in Ecuador fired the general secretary of one of the country's few banana unions when he demanded contract enforcement in 2001. In 2005, Dole obtained decertification of another agricultural union (USLEAP 2005, 5).

3. Violators spanned the range of Ecuadorian producers, including Chiquita (Human Rights Watch 2002).

4. See Juan Ferraro, "In Ecuador's Banana Fields, Child Labor Is Key to Profits," *New York Times*, July 13, 2002, 1; and "Leftist, a Former Coup Leader, Heads for Victory in Ecuador," *New York Times*, November 25, 2002, 8.

5. Although Noboa engaged a worker-rights omsbudsman to work with the new labor minister to alleviate union firings at his Primivera supplier (USLEAP 2005, 5), he did not dramatically change otherwise.

6. Human Rights Watch, petition on Ecuador's trade status to the USTR, 2005.

7. Dole's Josefa supplier had promised to reinstate eighty-six fired workers in mid-2005. Noboa/Bonita terminated purchases from San Luis (USLEAP, 2006a).

8. In October 2005, USLEAP demonstrated in front of the SAI meeting in New York. Despite having promised to negotiate, it took Dole until July 2008 to complete an agreement with SINTRASPLENDOR flower workers in Colombia (See USLEAP 2005, 2007).

9. In 1977, three-dozen workers were proved sterile; six won a $4.9 million judgment against Dow.

10. In 2002, a Nicaraguan court ordered Dole (jointly with Dow and Shell) to compensate 583 banana workers $490 million. When the case went to the United States, Dole countersued, alleging that the plaintiffs had falsified laboratory tests to impugn the company (*Dole, Behind the Smokescreen* 2006, 26).

11. Because of persistent sigatoka, Dole Europe's vice president for corporate relations does not believe organics could be a solution. "It is more promising in Ecuador and Peru. But currently, because of climate conditions, only 2 percent of bananas can be organic" (Cuperlier 2005).

12. FLO does not demand that "traders" like Dole recognize unions, but they must respect ILO Conventions 89 and 98.

13. *El Regional de Piura*, October 20, 2005. (Of the other two, BIOCOSTA sells to T.PORT in Germany; 99 percent of BIORGANIKA is owned by AgroFair.) (BanaFair 2007).

14. Group letter to Dole CEO David DeLorenzo, August 3, 2007 (copy also available from author).

15. See COLSIBA letter to Cuperlier, Dole, January 28, 2008; Banana Link letter, January 28, 2008. Copies also available from author.

Chapter 14. A Proposal for Fairer Trade

1. Myers's excellent study (2004) demonstrates Windward and U.K. efforts but may uncritically accept the inevitability of rationalized production.

2. Chiquita argued that although the EU's price-gap calculations and supply/demand equilibrium models claimed a 230-euro tariff would still allow market access, the actuality did not maintain market equilibrium (Stephen 2005). Chiquita offered other plausible econometric projections, citing Borrell and Bauer (2004) and its own later experience of losing $3/share in 2005-7 (Associated Press, February 9, 2008).

3. Del Monte claimed that TNCs had compensated for losses in the U.S. market with higher prices in Europe. However, quotas prevented Guatemalan shipments to Europe, and EU license requirements artificially inflated TNC prices, forcing companies out (Garcia 2005).

4. Compiled by Tearfund, Traidcraft, Christian Aid, ActionAid and CAFOD.

5. African participants included the East African Farmers Federation, Sub-regional Platform of Peasant Organizations of Central Africa, Network of Peasant and Agricultural Producers' Organizations in West Africa, Southern African Confederation of Agricultural Unions. In 2006 they met in the Dominican Republic.

6. Smith (2008) estimated the division would benefit the Dominican Republic, Belize, and certain West African importers such as Dole/Compagnie Fruitiére.

7. GSP+ contains evaluative mechanisms and complaint procedures similar to those of GSP in the United States. While government-based, they have stronger sanctions than ILO findings (see Frundt 1998 on the U.S. GSP).

8. While 75 percent of the tariff is assigned to the port of entry, EUROBAN proposed returning the remaining 25 percent to producer groups (Lombana and Parker 2006).

9. Group letter to EU Commission Chair José Sócrates Carvalho Pinto de Sousa, prime minister of Portugal, June 28, 2007 (copy also available from author).

Chapter 15. Toward a Sustainable Banana Alliance

1. This finding is consistent with the behavior of other farm-labor alliances such as Farmer-Labor Party success in Minnesota, 1920-40 (Valelly 1989).

2. This was my finding, allowing for Moberg's and Shreck's reservations (chap. 8).

3. In response to U.S. rank and file (e.g., the California Federation of Labor) and NGO complaints, the center has become more sensitive to local perceptions and worked to replace dependence on government funding.

4. USLEAP's early founders (then called U.S./Guatemalan Labor Education Project) included unionists from the Amalgamated Clothing Workers (now UNITE/HERE); United Electrical Workers; Health Care Workers Union 1199; the American Federation of State, County, and Municipal Employees; Service Employees International Union;

International Longshoremen's and Warehousemen's Union; and the IUF. Steven Coats, director since 1990, also developed strong working relationships with American Federation of Teachers locals, the International Union of Electrical Workers, RWDSU, UFCW, ITGLWF, COLSIBA, FESTRAS, FENACLE, UNTRAFLORES, and others.

5. Owing to USLEAP's ongoing vigilance, USTR kept Guatemala under review for five years, the most extensive worker-rights scrutiny given any nation under the GSP program. GSP pressure was critical in persuading the Guatemalan government to raise the minimum wage 50 percent, to establish new labor courts, and to gain legal, if short-lived, recognition of a dozen maquiladora unions.

6. These included the American Friends Service Committee, the Campaign for Labor Rights, Guatemalan Human Rights Commission/USA, Guatemala News and Information Bureau, Latin American Committee for Labor Advancement, Latin America Working Group, the Maquila Solidarity Network of Canada, the Network in Support of the People of Guatemala, the Nicaragua Network, the Religious Task Force on Central America and Mexico, Resource Center of the Americas, the Support Committee for Maquiladora Workers, the Washington Office on Latin America, Witness for Peace, and numerous local groups. In Latin America, USLEAP has also worked with the Center for Human Rights Legal Action, the Commission for the Verification of Codes of Conduct, and others.

7. Examples of these demonstrations include support actions for workers at Inexport, Confecciones Unidas, Phillips Van Heusen, and Starbucks suppliers in Guatemala. USLEAP helped to found STITCH to support women workers.

8. As I indicated in chapter 3, purchases of Fair Trade products grew in all European nations between 2001 and 2007, but discount chains have now entered the business, some attempting to control Fair Trade, others to undercut the approach. While surveys by Chiquita and Dole indicated consumer reluctance to pay substantially higher prices (interview with Vázquez 2004; Cuperlier 2005), FLO tests suggested they would be willing to expend 10 percent more for FT products (Ransom 2001, 95). This proved correct for bananas in 2007 (FLO 2007). In 2008, higher fuel prices brought a surcharge to both conventional and fair-trade bananas, but the fruit still remained a good consumer value.

9. In the United States the TransFair label could offer hope, given the success of union-friendly labels employed by agricultural unions in alliance with consumers, exemplified by United Farm Worker campaigns on strawberries, lettuce, and grapes (Frank 1999). Because of purchaser pressure, the Farm Labor Organizing Committee and the Immokalee Workers have achieved agreements with TNCs that handle tomatoes. FairTrade/union-labeled bananas could also become a popular mark since 65 percent of U.S. households buy the yellow fruit at least once a week, compared to only 31 percent that purchase apples. With bananas as profitable as they are, consumer pressure on supermarkets could be especially effective when combined with environmental demands.

Bibliography

Publications

Abbott, Diane. 2007. "What Is the Future for Jamaica's Banana Industry?" *Jamaica Observer*, April 22.

ACILS (American Center for International Labor Solidarity). 2006. *Justice for All: The Struggle for Worker Rights in Colombia*. Washington, D.C.: ACILS (May).

———. 2009. *Justice for All: The Struggle for Worker Rights in Guatemala*. Washington, D.C.: ACILS.

Action Aid. 2007. *Who Pays? How British Supermarkets are Keeping Women Workers in Poverty*. Action Aid, pdf file, http://www.actionaid.org.uk.

Acuerdo marco entre corporación de Desarrollo Bananero de Costa Rica S.A. BANDECO–Del Monte y El Sindicato de Trabajadores de Plantaciones Agrícolas (SITRAP). 1998. Siquirres, Costa Rica. (Febrero).

Adjei-Mensah, Simon. 2007. "Ghana Agricultural Workers Union (GAWU) Report." Presentation to EUROBAN, October 29, Geneva, Switzerland.

Alexander, Robin. 1999. "Experience and Reflections on the Use of the NAALC." *Memorias: Encuentro Trinacional de Laboralistas Democráticos*. México, D.F.: Universidad Nacional Autónoma.

Almanza-Alcalde, Horacio. 2005. "Transnational Social Movements, Solidarity Values and the Grassroots: The Fair Trade Movement, Mexican Coffee Production and a European NGO Coalition." Vinculado.org (Latin American electronic magazine on sustainable development), 31 Mayo.

Alvarez, Sonia, Evelyn Dagnino, Arturo Escobar, eds. 1998. *Cultures of Politics, Politics of Cultures*. Boulder, CO: Westview.

Amador, Armando. 1992. *Un Siglo de Lucha de los Trabajadores de Nicaragua (1880–1979)*. Managua: Universidad Centroamericana.

Amin, Samir. 1977. *Imperialism and Unequal Development*. New York: Monthly Review Press.

Anderson, Margaret L., and Howard F. Taylor. 2006. *Sociology: The Essentials*. 3rd. ed. Belmont, CA: Wadworth/Thomson Learning.

Andreatta, Susan L. 1998a. "Agrochemical Exposure and Farmworker Health in the Caribbean: A Local/Global Perspective." *Human Organization* 57, no. 3:350–58.

———. 1998b. "Transformations of the Agro-Food Sector: Lessons from the Caribbean." *Human Organization* 57, no. 4:414–29.

Arcury, Thomas A., Sara A. Quandt, Pamela Rao, Gregory B. Russell. 2001. "Farmworker Reports of Pesticide Safety and Sanitation in the Work Environment." *American Journal of Industrial Medicine* 39, no. 5:487–98.

Arguedas Mora, Carlos (Secretaria General, SUTAP). 1999a. "Avances en los Aspectos Ambientales en las Fincas Bananeras." *COLSIBA*, no. 10:4–5.

———. 2000. "Matanza de peces." (Febrero) http://members.tripod.com/foro_emaus.

———. 2002. "Importante logro ambiental en Costa Rica." http://www.colsiba.org.

Argueta, Mario. 1992. *Historia de los sin historia*. Tegucigalpa: Edi. Guaymuras.

———. 1995. *La Gran Huelga Bananera: Los 69 días que estremecieron a Honduras*. Tegucigalpa: Edi. Universitaria.

Armbruster, Ralph. 2005. *Globalization and Cross-Border Labor Solidarity in the Americas*. New York: Routledge.

Armstrong, Robert, and Janet Shenk. 1982. *El Salvador: The Face of Revolution*. Boston: South End Press.

Armstrong, Robert, Hank Frundt, Hobart Spalding, and Sean Sweeney. 1987. *Working Against Us: AIFLD and the International Policy of the AFL-CIO*. New York: NACLA.

Arnett, Allison. 2007. "The Banana's Appeal Extends to the Fair Trade Movement." *Boston Globe*, January 10.

Arreglo Directo entre Trabajadores de Finca Imperio Uno y BANDECO–Del Monte, S.A. 1999. Morales, Guatemala: SITRABI.

Astorga, E. Yamileth. 1998. "The Current Environmental Impact of the Banana Industry." Paper delivered at the International Banana Conference, Brussels, Belgium, May 4–6.

Asturias, Miguel Angel. 1976. *Viento Fuerte*. Buenos Aires, Argentina: Losada.

Ayres, Jeffrey M. 2002. "Transnational Political Processes and Contention Against the Global Economy." In Smith and Johnston 2002:191–206.

Bacon, David. 2004. *Children of NAFTA*. Berkeley and Los Angeles: University of California Press.

Ballinger, Jeff. 2007. "'Corporate Social Responsibility' Is Not Working for Workers . . . How Can We 'Anti-Sweat' Activists Have So Pathetically Little to Show for Fifteen Years of Cross-Border Organizing?" Faculty Seminar on Globalization, Labor, and Popular Struggles, Columbia University, New York, October 15.

Balzer, Deborah B. 1997. "The Impact of Environmental Factors on Factionalism and Schism in Social Movement Organizations." *Social Forces* 76, no. 1:199–228.

BanaFair. 2007. Report on CEPIBO, Peru (July).

BanaFair and Peuples Solidaire. 2006. Press release, July.

Banana Link. 2000. "Race to the Bottom: Banana Workers' Rights in Ecuador." Norfolk, U.K.: Banana Link.

———. 2001. *Banana Trade News Bulletin*, no. 24 (November).

———. 2003. "*Solidarismo*," or *Union-Busting Costa Rica–Style*. Norfolk, U.K.: Banana Link and War on Want.

———. 2005. "SITRAAMERIBI on Strike." *Banana Trade News Bulletin*.

———. 2007. *Asda/Walmart Price Cut Outrages Workers in Latin America*. Press release, April 16.
Banco de Guatemala. 1998. *Estadisticas*. Guatemala City: Banco de Guatemala.
Barlow, Maude, and Tony Clark. 2001. *Global Showdown*. Toronto: Stoddart.
Barrantes Cascante, Ramón (Sec. General SITAGH). 1999. "The aftermath of DBCP." Foro ed. II. http://members.tripod.com/foro_emaus/complete Foro2000.htm:2–5.
Barrantes Cascante, Ramón, and Padre Gerardo Vargas Varela (Director, Foro Emaús). 2000. "Denuncia ante la opinión pública . . . ," 17 Julio. See http://members.tripod.com/foro_emaus.
Barratt Brown, Michael. 1993. *Fair Trade: Reform and Realities in the International Trading System*. London: Zed Books.
Bartlett, Donald L., and James B. Steele. 2000. "How to Become a Top Top Banana." *Time*, February 7, 42–56.
Bastian, Hope. 2006. "Keeping Fair Trade Fair in Mexico," *NACLA* 39, no. 6:6–9.
Baughen, Cameron. 2003. "What are the Limits of Fair Trade?" *Alternatives*, December 15.
Beck, Ulrich. 1992. *The Risk Society*. London: Sage.
Bendell, Jem, ed. 2000. *Terms for Endearment: Business, NGOs and Sustainable Development*. London: Greenleaf Publishers.
Bermúdez Umaña, Gilberth (Coord., COLSIBA). 1998. "Repression in the Atlantic Zone of Costa Rica" in Foro Emaús 1998d, 14–18.
Black, Jan Kippners. 1999. *Development in Theory and Practice*. 2nd ed. Boulder, CO: Westview Press.
Bolland, O. Nigel 1992. "The Politics of Freedom in the British Caribbean." In *The Meaning of Freedom*, ed. by Frank McGlynn and Seymour Drescher. Pittsburgh: University of Pittsburgh Press.
Booth, John. 1998. *Costa Rica: Quest for Democracy*. Boulder, CO: Westview.
Bourdieu, Pierre. 1992. *The Logic of Practice*. Stanford: Stanford University Press.
Borrell, Brent, and Marcia Bauer. 2004. *EU Banana Drama: Not Over Yet*. Canberra, Australia: Center for International Economics (March).
Boshart, Jeff. 2004. "Biodynamic Farming Pioneers Revolutionize Banana Production in the Dominican Republic." *The New Farm Regenerative Agriculture Worldwide* (Rodale), April 6.
Bourgois, Philippe. 1989. *Ethnicity at Work*. Baltimore: Johns Hopkins University Press.
Brandy, Joe, and Jackie Smith, eds. 2005. *Coalitions Across Borders*. Lanham, MD: Rowman & Littlefield.
Broad, Robin. 2002. *Global Backlash*. Lanham, MD: Rowman & Littlefield.
Brodzinsky, Sibylla. 2007. "Chiquita Case Puts Big Firms on Notice." *Christian Science Monitor*, April 11.
Bros, Warner. 1999. *El Libro de la Banana*. Planeta Publishing Corp.
Brosius, J. Peter. 1999. "Green Dots, Pink Hearts: Displacing Politics from the Malaysian Rain Forest." *American Anthropologist* 101, no. 1:36–57.

Buchanan, Millie, and Gerry Scoppettuolo. 1997. "Environment for Cooperation: Building Worker Community Coalitions." In *Work, Health and Environment*, ed. by Charles Levenstein and John Wooding, 329–51. New York: Guilford Press.

Bucheli, Marcelo. 1997. "United Fruit Company in Colombia: Impact of Labor Relations and Governmental Regulations on Its Operations, 1948–1968." *Essays in Economic and Business History* 15:65–84.

———. 2005. *Bananas and Business: The United Fruit Company in Colombia, 1899–2000*. New York: New York University Press.

Buitrago, Gelkha (FLO). 2007. *Is There a Need for an ISO Standard on Fair Trade?* Submission of the International Fair Trade Movement to the Committee on Consumer Policy of the International Organization for Standardization for the Twenty-ninth Plenary Meeting in Salvador Bahia, Brazil. Brussels: Fair Trade Advocacy Office, May.

Burbach, Roger. Forthcoming. *Fractured Utopias*.

Burbach, Roger, and Patricia Flynn. 1980. *Agribusiness in the Americas*. New York: Monthly Review Press.

Carty, Bob. 2001. "Fair Trade Broadcast Transcript." *Living on Earth*, transcript of radio presentation, Nov. 30.

CBS 60 Minutes. 2008. "The Price of Bananas." Produced by Andy Court, May 11.

Center for Environmental Studies. 1996. "Banana Production in Costa Rica." *Reading Materials*, 71–94. London: Lead International.

CEPIBO. 2006. Presentation to EUROBAN, September 14.

Chambron, Anna. 2000. *Straightening the Bent World of Bananas*. Maastricht: EFTA.

———. 2005. "Can Voluntary Standards Provide Solutions?" In IBCII Preparatory Papers, 94–107.

Chapman, Peter. 2008. *Bananas!: How The United Fruit Company Shaped the World*. New York: Canongate.

Chiquita Brands. 2000. *Corporate Responsibility Report*. Cincinnati: Chiquita Brands.

———. 2001. *Corporate Responsibility Report*. Cincinnati: Chiquita Brands.

———. 2007a. *Plea Agreement with Dept. of Justice*, March 14.

———. 2007b. *Annual Report*.

Chomsky, Aviva. 1996. *West Indian Workers and the United Fruit Company in Costa Rica, 1870–1940*. Baton Rouge: Louisiana State University Press.

———. 2004a. "Labor and Violence in Colombia: A Global History." Paper delivered at Northeastern University, World History Center panel, Boston, March 13.

———. 2004b. "What's Old, What's New? Globalization, Violence and Identities in Colombia's Banana Zone." Presented at LASA Miniconference on "Labor and Globalization," Las Vegas, October 6.

Chomsky, Aviva, and Aldo Lauria-Santiago, eds. 1998. *Identity and Struggle at the Margins of the Nation-State*. Durham, NC: Duke University Press.

CLAC. 2005. *Boletin*, no. 1, December 16.

———. 2006a. *Boletin*, no. 5, August 20.

———. 2006b. *Segunda CLAC Asemblea General*, Dominican Republic, October 26–28.

Clinton, William J. 1999. *Presidents Chirac-Clinton Press Conference*. December 19.

Coats, Stephen (Dir., USLEAP). 1998. *Reflections on the Issue of Independent Monitoring.* Washington D.C.: Campaign for Labor Rights.
Cockcroft, James, Andre Gunder Frank, and Dale Johnson. 1971. *Dependence and Underdevelopment: Latin America's Political Economy.* Garden City, NY: Doubleday.
Collier, Ruth Berins, and David Collier. 1991. *Shaping the Political Arena.* Princeton: Princeton University Press.
COLSIBA. 1998. "Negociaciones con Chiquita." *COLSIBA,* no. 7:2–3.
———. 1999. "COLSIBA Propone firma de Acuerdo Regional de Carácter Laboral y Ambiental." *COLSIBA,* no. 9:2.
———. 2001. "Sindicalismo Bananero Latinoamericano ante los Códigos de Conducta e Iniciativas de Responsabilidad Social Voluntaria." Presentation at the Conference on Voluntary Socio-Labor Standards in the Banana Industry, London, September 27–28.
———. 2007. *Actualización de la situación con la empresa Dole en la region.* Julio.
COLSIBA, EUROBAN, and WINFA. 2006. *Monitoring Impacts of the New EU Banana Import Regime.* August.
Compa, Lance. 2001. "Wary Allies." *American Prospect* 12, no. 2 (July):2–16.
Compa, Lance, and Stephen F. Diamond. 1996. *Human Rights, Labor Rights and International Trade.* Philadelphia: University of Pennsylvania Press.
Conroy, Michael E. 2006. "Transnational Social Movements Linking North and South: The Struggle for Fair Trade." Presented at "Alternative Visions of Development" conference, University of Florida, Gainesville, February 23–25.
———. 2007. *Branded! How the Certification Revolution Is Transforming Global Corporations.* Gabriola Island, BC, Canada: New Society Publishers.
Cooper, Peter, and Lori Wallach. 1995. *Nafta's Broken Promises.* Washington, D.C.: Public Citizen.
Coor, Anders. 1999. "Battling the Banana Baron: Rural Hondurans Bloody Chiquita Brands International." In *No Trespassing: Squatting, Rent Strikes and Land Struggles Worldwide,* 39–50, Boston: South End Press.
CORBANA (Corporación Bananera Nacional). 1999. Dirección de Política Bananera y Estadística: San José: Informe anual de estadísticas de exportación de banano, 1996–1998.
———. 2004. *Informe Anual de Labores, 2003.* San José: CORBANA.
COSIBAH. 2005. *Boletin.* (4–15)
COSIBA-CR. 1997. *Cláusulas Sociales y Ambientales en la Producción Bananera.* San José: Costa Rican Banana Unions.
———. 1999. *Propuesta de la coordinadora de sindicatos bananeros de Costa Rica sobre la problemática de las libertades sindicales y la negociación colectiva en las plantaciones bananeras de Costa Rica.* San José: Costa Rican Banana Unions. May 25.
———. 2000. *Chiquita niega derecho a negociación colectiva y desmejora condiciones laborales y salariales.* San José: Costa Rican Banana Unions. See http://members.tripod.com/foro_emaus.
Crichlow Michaeline A. 2003. "Neoliberalism, States, and Bananas in the Windward Islands." *Latin American Perspectives* 30, no. 3, (May):37–57.

Cuperlier, Sylvain. 2005. "Race to the Bottom: Dole's Perspective." IBCII Preparatory Conference, Washington, D.C., February 2.

———. 2007. Cuperlier to B. Peyrot des Gachons, Peuples Solidaires, August 14.

Cutter, Victor M. (UF president). 1926. "Caribbean Tropics in Commercial Transition." *Economic Geography* 2, no. 4:494–507.

Danaher, Kevin. 1994. *Fifty Years Is Enough: The Case Against the World Bank and the International Monetary Fund.* Boston: South End Press.

della Porta, Donatella, and Sydney Tarrow. 2005. *Transnational Protest and Global Activism.* Lanham, MD: Rowman & Littlefield.

Dickenson, Rink, and Rob Everts (Co-Executive Directors, Equal Exchange). 2005. Dickinson and Everts to FLO International Board of Directors, June 22.

Diller, Janelle. 1999. "A Social Conscience in the Global Marketplace? Labour Dimensions of Codes of Conduct, Social Labeling and Investor Initiatives." *International Labour Review* 138, no. 2:9–129.

Dillon, Tim. 2004. "The Fair Trade Issue Is About to Percolate from the Coffee Pot to the Fruit Bowl." *USA Today,* January 20.

Divney, Tom (Rainforest Alliance). 2005. "Environmental and Social Certification." IBCII Preparatory Conference, Washington, D.C., February 2.

Dole, Behind the Smokescreen. 2006. Written and produced by various NGOs and unions; pdf available at www.usleap.org/files.

Dosal, Paul J. 1993. *Doing Business with the Dictators: A Political History of United Fruit in Guatemala, 1899–1944.* Wilmington, DE: Scholarly Resources.

Dreiling, Michael, and Ian Robinson. 1998. "Union Responses to NAFTA in the U.S. and Canada: Explaining Intra- and International Variation." *Mobilization* 3, no. 2:163–84.

Dublin-Prince, Josephine. 2005. "Women in the Windward Islands." *IBCII Preparatory Papers,* 90–94.

Dunning, John H. 1998. "The Origins of Convention 87 on Freedom of Association and the Right to Organize." *International Labour Review* 137, no. 2:149–68.

Edelman, Mark. 1999. *Peasants Against Globalization.* Stanford, CA: Stanford University Press.

Emmanuel, Arghiri, and Brian Pearce, eds. 1972. *Unequal Exchange: A Study of Imperialism and Trade.* New York: Monthly Review Press.

Equal Exchange. 2005. *Fair Trade Coffee Pioneer Questions Nestlé's Entry into Market.* Press release, October 7.

Ericson, Rose Benz. 2006. *The Conscious Consumer: Promoting Economic Justice Through Fair Trade.* Madison, WI: SERRV International.

Esbenshade, Jill. 2004. *Monitoring Sweatshops.* Philadelphia: Temple University Press.

Escobar, Arturo, and Sonia Alvarez. 1992. *The Making of Social Movements in Latin America.* Boulder, CO: Westview Press.

Escofet, Guillermo. 1998. "Female Banana Workers at Risk." *Tico Times,* May 29.

Ethibel. 2004. "Stock at Stake in Banana Trade." *News Bulletin,* 30–31.

Euraque, Dario A. 1996. *Reinterpreting the Banana Republic: Region and State in Honduras, 1870–1972.* Chapel Hill: University of North Carolina Press.

EUROBAN. 1997. "The Message of Solidarity of EUROBAN: Our Responsibility for the Problems around the Banana Industry." In *Foro Emaus,* 1998d.

———. 2006. *Monitoring Impacts of the New EU Banana Import Regime.* (August).

———. 2007. EUROBAN to Dole CEO David Murdock, July 19.

Eurofruit (The international magazine of fresh produce). 2002. "Costa Rica Adjusts to New Market Realities." London. June 28.

Evans, Michael. 2007. "'Para-Politics' Goes Bananas." *The Nation* (Web), April 4. http://www.thenation.com/doc/20070416/evans.

Fairclough, Gordon, and Darren McDermott. 1999. "The Banana Business Is Rotten, So Why Do People Fight Over It?" *Wall Street Journal,* August 9:1.

Fallas, Carlos. 1975. *Mamita yunai.* San José, Costa Rica: Lehman S.A.

FAO. 1999. *The Market for 'Organic' and 'Fair-trade' Bananas.* Gold Coast, Australia: FAO Committee on Commodity Problems, Intergovernmental Group on Bananas and Tropical Fruits, May 4–8.

———. 2003. *Organic and Fair Trade Bananas and Environmental and Social Certification in the Banana Sector.* (CCP:BA/TF 03/5). Rome: FAO. December.

FAO/ITC/CTA. 2001. *World Market for Organic Fruits and Vegetables.* Rome: FAO/ITC/CTA.FAOSTAT. 2005–7. www.faostat.fao.org.

Featherstone, Lisa. 2002. *Students Against Sweatshops: The Making of a Movement.* London: Verso.

Fireman, Bruce, and William Gamson. 1979. "Utilitarian Logic in the Resource Mobilization Perspective." In *The Dynamics of Social Movements,* ed. by Mayer N. Zald and John D. McCarthy, 6–34. Cambridge: Winthrop Publishers.

Fischer, Helge. 2002. "Tendencias recientes del mercado mundial del banano." *Revista Foro* (Costa Rica), 2:32–36.

———. 2007a. *Urgent Action About Intoxications and Dismissal of Workers on the Banana Farm Coyol, Chiquita/Cobal, Costa Rica, and Review of the Chiquita Campaign.* BanaFair, May 6.

———. 2007b. *Dole Campaign Resume.* BanaFair, June 7.

FLO (Fairtrade Labelling Organization International). 2004. *Standards.* Bonn: FLO.

———. 2005. *Fairtrade Producer Profile: Grupo Agricola Prieto (Ecuador).* January. Pdf file available from author.

———. 2006a. *Fairtrade Producer Profile: ABBPOSA (Peru).* March. Pdf file available from author.

———. 2006b. *Fairtrade Producer Profile: APOQ (Peru).* March. Pdf file available from author.

———. 2006c. *Annual Report.* Bonn: FLO.

———. 2007. *Annual Report.* Bonn: FLO.

FLO/IUF. 2007. Conference, "Summary: FLO and Trade Unions—Working Together for Fair Labour Conditions." Bonn, Germany, June 25–26.

Flores Castillo, Warren. 1999. *La realidad de los derechos laborales de los trabajadores del sector bananero de Costa Rica*. San José, Costa Rica: COSIBACR/OIT.

Flores Madrigal, Juan José. 1993. *El Solidarismo desde Adentro*. 3rd ed. San José, Costa Rica: ASEPROLA.

Foro Emaús. 1998a. *Bananas for the World and the Negative Consequences for Costa Rica?* San José, Costa Rica: Foro Emaús.

———. 1998b. "Entre la vergüenza y la dignidad." Sup. esp., *La Voz Del Manatí* (Costa Rica), Diciembre.

———. 1998c. *Normas mínimas socio-ambientales para la producción bananera en Costa Rica*. San José, Costa Rica: Foro Emaús. http://members.tripod.com/foro_emaus.

———. 1998d. *Foro: The Secret Accounts of a Banana Enclave*. San José, Costa Rica: Foro Emaús.

———. 1999a. *Reporte anual*.

———. 1999b. Foro Ed. II. See http://members.tripod.com/foro_emaus.

———. 2000. *DBCP en la Producción Bananera*. See http://members.tripod.com/foro.

———. 2001. *Por la prohibición del Paraquat en C.R: Informe Especial*. (Marzo).

———. 2003. *10 años de lucha*. San José, Costa Rica: Foro Emaús.

Foro Emaús and the International Union of Foodworkers. 2001. *El uso de plaguicidas y su relación con el desarrollo en Costa Rica*. (Noviembre).

Foweraker, Joe. 1995. *Theorizing Social Movements*. London: Pluto Press.

Frank, Dana. 1999. *Buy American: The Untold Story of Economic Nationalism*. Boston: Beacon Press.

———. 2002. "Our Fruit, Their Labor, and Global Reality." *Washington Post*, June 2.

———. 2005. *Bananeras: Women Transforming the Banana Unions of Latin America*. Boston: South End Press.

Freeman, Jo. 1983. *Social Movements of the Sixties and Seventies*. New York: Longman's.

French, John. 1994. "The Declaration of Philadelphia and the Global Social Charter of the U.N., 1944–45." In *International Labour Standards and Economic Interdependence*, ed. by Werner Sengenberger and Duncan Campbell, 19–28. Geneva: ILO.

Fridell, Gavin. 2006. "Fair Trade and Neoliberalism: Assessing Emerging Perspectives." *Latin American Perspectives* 33, no. 6:8–28.

FruitTrop Focus (Annual French Editions). 1992.

———. 1998.

———. 1999.

Frundt, Henry J. 1995. "The Rise and Fall of AIFLD in Guatemala." *Social and Economic Studies* 44, nos. 2 and 3:287–319.

———. 1998. *Trade Conditions and Labor Rights: U.S. Initiatives, Dominican and Central American Responses*. Gainesville, FL: University Press of Florida.

———. 2003. *Pausas Refrescantes*. Rev. ed. Guatemala City: Magna Terra.

———. 2004a. "Unions Wrestle with Corporate Codes of Conduct." *Working USA* 7, no. 4:36–69.

———. 2004b. "Ending Central American Quotas: How Global Firms and Local Unions Respond." LASA presentation, Las Vegas, NV, Oct. 9.

———. 2005a. "Hegemonic Resolution in the Banana Trade." *International Political Science Review* 26, no. 2:215–37.
———. 2005b. "Movement Theory and International Labor Solidarity." *Labor Studies Journal* 30, no. 1:19–40.
Frundt, Henry J., and Chinchilla, Norma. 1987. "Trade Unions in Guatemala." In *Latin American Labor Organizations*, ed. by Gerald. M. Greenfield and Sheldon L. Marman, New York: Greenfield Press.
FTF (Fair Trade Foundation, U.K.). 2004a. *Introducing Fair Trade*. Fairtrade Resources FM1, April.
———. 2004b. *Fairtrade Bananas—Looking Behind the Price Tag*.
———. 2006. www.fairtrade.org.uk/resources/reports.
———. 2007. "Luis, El Guabo." www.fairtrade.org.uk/resources/reports.
———. 2008. Statement on Supermarket Banana Price Cuts. May.
Furtado, Celso. 1976. *Economic Development of Latin America*. 2nd ed. London: Cambridge University Press.
Gacek, Stan. 2005. "A Rejoinder to 'Revolution and Counter-Revolution.'" *New Labor Forum* 14, no. 3.
Galeano, Eduardo. 1969. *The Open Veins of Latin America*. New York: Monthly Review Press.
Gallagher, Mike, and Cameron McWhirter. 1998. "Chiquita Secrets Revealed: Hidden Control Crucial to Overseas Empire." *Cincinnati Enquirer*, May 3:S1–S18.
Garcia, Marco (Del Monte). 2005. "Race to the Bottom: Del Monte's Perspective." North American Preparatory for International Banana Conference II, Washington, D.C., February 2.
Gearhart, Judy (SAI). 2005. "The Social Accountability Initiative." North American Preparatory for International Banana Conference II, Washington, D.C., February 2.
Gerbarh, Jurgen, and Dieter Rucht. 1993. "Mesomobilization: Organizing and Framing in Two Protest Campaigns in West Germany." *American Journal of Sociology* 98, no. 3:555–95.
Gernigon, Bernard, Alberto Odero, and Horatio Guido. 2000. "ILO Principles Concerning Collective Bargaining." *International Labour Review* 139, no. 1:33–56.
Getz, Christy, and Aimee Shreck. 2006. "What Organic and Fair Trade Labels Do Not Tell Us: Towards a Place-based Understanding of Certification." *International Journal of Consumer Studies* 30, no. 5:490–501.
Giugni, Marco, Doug McAdam, and Charles Tilly. 1999. *How Social Movements Matter*. Minneapolis: University of Minnesota Press.
Glazer, Barney G., and Anselm A. Strauss. 1967. *The Discovery of Grounded Theory*. Chicago: Aldine.
Gleijeses, Piero. 1991. *Shattered Hope*. Princeton, NJ: Princeton University Press.
Godfrey, Claire. 1998. *A Future for Caribbean Bananas*. Oxfam Briefing Paper, Oxfam U.K. Policy Dept. March.
Goffman, Erving. 1997. "Frame Analysis." In *The Goffman Reader*, ed. by Charles Lemert and Ann Brananman, 14–166. Malden, MA: Blackwell.

González, Mario Aníbal. 1988. "Algunos apuntes sobre la actividad productora de banano en Guatemala y las organizaciones sindicales." In *Cambio y continuidad en la economía bananera*, 185–202. San José, Costa Rica: FLACSO/CEDAL/FES.

Goulet, Denis. 1989. *Incentives for Development: The Key to Equity*. New York: New Horizons Press.

Grant, Clifton. 2005. "A Better Place to Work: A Jamaican Perspective." Presentation at the North American Preparatory for International Banana Conference II, Washington, D.C., February 2.

Greenfield, Gerald Michael, and Sheldon L. Maram, eds. 1987. *Latin American Labor Organizations*. Westport, CT: Greenwood Press.

Grimes, Kimberly M., and B. Lynne Milgram, eds. 2000. *Artisans and Cooperatives: Developing Alternative Trade for the Global Economy*. Tucson: Univ. of Arizona Press.

Grossman, Lawrence. 1998. *The Political Ecology of Bananas*. Chapel Hill: University of North Carolina Press.

———. 2003. "The St. Vincent Banana Growers' Association, Contract Farming and the Peasantry." In Striffler and Moberg, 286–315.

Habermas, Jürgen. 1984. *The Theory of Communicative Action*. Vols. 1 and 2. Boston: Beacon Press.

———. 1989. *The Structural Transformation of the Public Sphere*. Cambridge: MIT Press.

Harari, Raul. 2005a. "The Environmental and Health Impacts of Banana Production in Latin America." *IBCII Preparatory Papers*, 40–56.

———. 2005b. "The Working and Living Conditions of Banana Workers in Latin America." *IBCII Preparatory Papers*, 57–85.

Haufler, Virginia. 2001. *A Public Role for the Private Sector*. Washington, D.C.: Carnegie Endowment for International Peace.

Hellin, Jon, and Sophie Higman. 2001. "The Impact of the Multinational Companies on the Banana Sector in Ecuador." Oxford: hellin@fincahead.com.

Henriques, William, Russel D. Jeffers, Thomas E. Lacher, Jr., and Ronald J. Kendall. 1997. "Agrochemical Use on Banana Plantations in Latin America: Perspectives on Ecological Risk." *Environmental Toxicology and Chemistry* 16:91–99.

Hermosilla, Hernán (Vice Director, Foro Emaús). 1998. "The History of the Foro Emaús: A Grass Roots and Ecumenical Struggle for the Defense of Life." In *The Secret Account of a Banana Enclave*, 8–13. San José, Costa Rica: Foro Emaús.

———. 2000. "Trabajadores bananeros despedidos . . . por Finca Bananera Dos Rios, S.A." (Abril-Julio). San José: Foro Emaús. See http://members.tripod.com/foro_emaus.

Hernández, Carlos E., and Scott G. Witter. 1996. "Evaluating and Managing the Environmental Impact of Banana Production in Costa Rica: A Systems Approach." *Ambio* 25, no. 3:171–78.

Hertel, Shareen, and Lance Minkler. 2007. *Economic Rights: Conceptual, Measurement, and Policy Issues*. New York: Cambridge University Press.

Herzenberg Stephen, and Jorge Perez-Lopez, eds. 1990. *Labor Standards and Development in the Global Economy*. Washington D.C.: U.S. Department of Labor.

Hipsher, Patricia L. 1997. "Democratic Transitions as Protest Cycles." In *The Social Movement Society*, ed. by David Meyer and Sydney Tarrow. Lanham, MD: Rowman & Littlefield.

Hirsch, J., and E. Aguilar. 1996. "Banana Expansion in the Humid Tropical Zones." in *Reading Material*, Training Session in Costa Rica, 95–108. London: LEAD (Leadership for Environment and Development) International.

Holderness, Mark, Suzanne Sharrock, Emile A. Frison, Moses Kairo, eds. 1999. *Organic Banana 2000: Toward an Organic Banana Initiative in the Caribbean*. Report of an international workshop. Dominican Republic: INABAB.

Holl, Karen D., Gretchen C. Daily, and Paul R. Ehrlich. 1995. "Knowledge and Perceptions in Costa Rica Regarding Environment, Population, and Biodiversity Issues." *Conservation Biology* 9, no. 6:1548–55.

Holt-Giménez, Eric, Ian Bailey, and Devon Sampson. 2007. "Fair to the Last Drop: The Corporate Challenges to Fair Trade Coffee." *Food First Development Rpt*. 17, June 2. San Francisco: Institute for Food and Development Policy.

Hopkins, Michael. 1998. *The Planetary Bargain: Corporate Social Responsibility Comes of Age*. New York: St. Martin's Press.

Human Rights Watch. 2002. *Tainted Harvest*. New York: Human Rights Watch.

Hurst, Peter. 2007. *Agricultural Workers and Their Contribution to Sustainable Agriculture and Rural Development*. Switzerland: IUF/ILO/FAO.

IBCII (International Banana Conference II). 2005a. Preparatory Papers. Brussels, April 28–30.

———. 2005b. "Reversing the Race to the Bottom: Final Conference Statement." Brussels, April 28–30.

IISD (International Institute for Sustainable Development). 1996. *Global Green Standards: ISO 14000 and Sustainable Development*. Winnepeg, Canada: IISD.

Immerman, Richard. H. 1982. *The CIA in Guatemala*. Austin: University of Texas Press.

Independent University Initiative. 2000. *Final Report*. Cambridge, MA: Business for Social Responsibility Education Fund, Investor Responsibility Research Center.

International Books. 1994. *Second International Water Tribunal*. Netherlands: International Books.

ISCOD (Instituto Sindical de Cooperación al Desarrollo)/Agencia Española de Cooperación Internacional. Diciembre 2003–Diciembre 2004. *Proyecto de Apoyo de Los Sindicatos Bananeros de Costa Rica* (documentos de taller: Propuestos de certificación ind. género; Códigos de Conducto; Negociación Colectiva y Conflicto de Carácter Económico y Social; Desarrollo y perspectivas de las organizaciones sociales de C.R.).

Jaffee, Daniel. 2007. *Brewing Justice*. Berkeley: University of California Press.

Jenkins, Virginia Scott. 2000. *Bananas: An American History*. Washington, D.C.: Smithsonian Press.

Jonas, Susanne. 1991. *The Battle for Guatemala*. Boulder, CO: Westview Press.

———. 2000. *Of Centaurs and Doves: Guatemala's Peace Process*. Boulder, CO: Westview Press.

Joseph, Gilbert M., Catherine C. LeGrand, and Ricardo D. Salvatore, eds. 1998. *Close Encounters of Empire*. Durham, N.C.: Duke Univ. Press.
Kamel, Rachael, and Anya Hoffman, eds. 1999. *The Maquiladora Reader*. Philadelphia: AFSC.
Karnes, Thomas L. 1978. *Tropical Enterprise: The Standard Fruit and Steamship Company in Latin America*. Baton Rouge: Louisiana State University Press.
Kazis, Richard, and Richard L. Grossman. 1982. *Fear at Work*. New York: Pilgrim Press.
Keck, Margaret, and Katherine Sikkink. 1998. *Activists Beyond Borders*. Ithaca N.Y.: Cornell University Press.
Kelly, Christine. 2001. *Tangled Up in Red, White and Blue*. New York: Rowman & Littlefield.
Kepner, Charles D. 1936. *Social Aspects of the Banana Empire*. New York: Columbia University Press.
Kepner, Charles D., and Jay Soothill. 1935. *The Banana Empire: A Case Study in Economic Imperialism*. New York: Russell and Russell.
Kinzer, Stephen, and Stephen Schlesinger. 1982. *Bitter Fruit: The Untold Story of the American Coup in Guatemala*. New York: Doubleday.
Klandermans, Bert. 1990. "Linking the 'Old' and the 'New': Movement Networks in the Netherlands." In *Challenging the Political Order*, ed. by Russell J. Dalton and Manfred Kuechler. New York: Oxford.
Klein, Naomi. 2000. *No Space, No Choice, No Jobs, No Logo*. New York: Picador.
Koeppel, Dan. 2007. *Banana: The Fate of the Fruit That Changed the World*. New York: Hudson Street Press.
Krier, Jean Marie. *Fair Trade in Europe 2005*. Brussels: FINE.
Krut, Riva, and Harris Gleckman. 1998. *ISO 14001: A Missed Opportunity for Sustainable Development*. London: Earthscan Publications.
Lamb, Harriet. 2008. *Fighting the Banana Wars and Other Fair Trade Battles*. London: Rider Books.
Langley, Lester D., and Thomas David Schoonover. 1995. *The Banana Men: American Mercenaries and Entrepreneurs in Central America, 1880–1930*. Lexington: University Press of Kentucky.
Lappé, Francis Moore, and Joseph Collins. 1975. *Food First*. New York: Ballantine.
Larrea Maldonado, Carlos. 1987. *El Banano en el Ecuador*. Quito: Corporación Edi. Nacional.
La Voz de la Mantí (newsletter). 1999. (Costa Rica). Agusto.
Lee, Charles. 1992. "Beyond Toxic Wastes and Race." In *Confronting Environmental Racism*, ed. by Robert. D. Bullard, 41–52. Boston: South End Press.
Leipziger, Deborah. 2001. *SA8000: The Definitive Guide to the New Social Standard*. London: Financial Times/Prentice Hall.
Levenson-Estrada, Deborah. 1994. *Trade Unions Against Terror*. Durham: University of North Carolina Press.
Lewis, Neil A. 2007. "Inquiry Threatens Ex-Leader of Securities Agency." *New York Times*, August 16:A19.

Liddell, Ian. 2000. *Unpeeling the Banana Trade*. Fair Trade Foundation. August.
Littrell, Mary A., and Marsha A. Dickson. 1997. "Alternative Trading Organizations: Shifting Paradigm in a Culture of Social Responsibility." *Human Organization* 56, no. 3:344–52.
Liu, Pascal, Michel Andersen, Catherine Pazderka. 2004. *Voluntary Standards and Certification for Environmentally and Socially Responsible Agricultural Production and Trade*. FAO Commodities and Trade Technical Paper No. 5. Rome: FAO.
Lombana, Jahir, and Liz Parker. 2006. *Where Next with Trade Policy?* EUROBAN Discussion Paper, September.
Longley, Susan (IUF). 2005. "Presentation on Global Agriculture and Occupational Health." North American Preparatory for International Banana Conference II, Washington, D.C., February 2.
Macdonald, Laura. 2005. "Gendering Transnational Social Movement Analysis: Women's Groups Contest Free Trade in the Americas." In Brandy and Smith 2005:21–42.
Magdoff, Fred, John Bellamy Foster, and Fredrich Buttel, eds. 2000. *Hungry for Profit: The Agribusiness Threat to Farmers, Food, and the Environment*. New York: Monthly Review Press.
Marquardt, Steve. 2002. "Pesticides, Parakeets, and Unions in the Costa Rican Banana Industry, 1938–1962." *Latin American Research Review* 37, no. 2:3–33.
Masibay, Kimberly. 2000. "Growers Tune: Yes We Have Ecobananas." *Christian Science Monitor*, July 19, 1–3.
Mattsson, Eva. 2004. *Comparison of Standards for Sustainable Cultivation of Bananas*. Stockholm: Swedish Society for Nature Conservation.
May, Roy. 1998. "Transnational Companies and Governments Against the People: The Struggle of Sara de Bataan." In Foro Emaús 1998d.
McAdam, Doug. 1982. *Political Process and the Development of Black Insurgency*. Chicago: University of Chicago Press.
McAdam, Doug, John D. McCarthy, and Mayer N. Zald, eds. 1996. "Introduction: Opportunities, Mobilizing Structures and Framing Processes—Toward a Synthetic, Comparative Perspective on Social Movements." In *Comparative Perspectives on Social Movements*, 1–20. Cambridge: Cambridge University Press.
McAdam, Doug, Sydney Tarrow, and Charles Tilly. 2001. *Dynamics of Contention*. New York: Cambridge University Press.
McCammon, Holly J., Karen E. Campbell, Ellen M. Granberg, and Christine Mowery. 2001. "How Movements Win: Gendered Opportunity Structures and U.S. Women's Suffrage Movements, 1866–1919." *American Sociological Review* 66, no 1:49–70.
McCann, Thomas P. 1976. *An American Company: The Tragedy of United Fruit*. New York: Crown Publishers.
McCarthy, John D., and Mayer N. Zald. 1973. *The Trend of Social Movements in America*. Morristown, NJ: General Learning Press.
McCracken, Carrie. 2002. *Sembrando sostenibilidad en Talamanca*. Siquirres, Costa Rica: Foro Emaús.
McLaughlin, David. (Chiquita Officer). 2006. Phone conversation with author, August.

Melucci, Alberto. 1989. *Nomads of the Present*. Philadelphia: Temple University Press.
——. 1996. *Challenging Codes*. Cambridge: Cambridge University Press.
Meza, Victor. 1997. *Historia del Movimiento Obrero Hondureño*. Tegucigalpa: Centro de Documentación de Honduras.
Meyer, David S. 2002. "Opportunities and Identities: Bridge Building in the Study Of Social Movements." In *Social Movements: Identity, Culture and the State*, ed. by David. S. Meyer, Nancy Whittier, and Belinda Robnett, 3–24. New York: Oxford University Press.
Meyer, David S., Nancy Whittier, and Belinda Robnett, eds. 2002. *Social Movements: Identity, Culture and the State*. New York: Oxford University Press.
Miller, Michael (Chiquita Brands). 2007. *Statement to Panamanian Press*, quoted by SOPISCO, August. A press conference attended by SOPISCO.
Ministerio de Salud, Costa Rica. 1998. *Diagnostico de Aspectos Ocupacionales y Ambientales de la Actividad bananera de la zona Atlántica de Costa Rica—1997*.
Moberg, David. 2000. "For Unions, Green's Not Easy." *The Nation*, February 21:17–21.
Moberg, Mark. 2005. "Fair Trade and Eastern Caribbean Banana Farmers: Rhetoric and Reality in the Anti-Globalization Movement." *Human Organization* 64, no. 1:4–15.
Mortensen, S., Kevin Johnson, C. Weisskopf, M. Hooper, T. Lacher, and R. Kendall. 1998. "Avian Exposure to Pesticides in Costa Rican Banana Plantations." *Bulletin of Environmental Contamination and Toxicology* 60:562–68.
MSN (Maquila Solidarity Network). 2001. *SA 8000: Can Commercial Auditing Promote Worker Rights?* Toronto, Report no. 8.
——. 2002. *Are Apparel Manufacturers Getting a Bad WRAP?* Toronto, Report no. 12.
Munguía, Iris. 2005. "For the Respect of Labour and Social Rights of Women Banana Workers in Latin America." *IBCII Preparatory Papers*, 86–90.
Murray, Douglas L. 1994. *Cultivating Crisis: The Human Cost of Pesticides in Latin America*. Austin: University of Texas Press.
Murray, Douglas, and Laura Raynolds. 2000. "Alternative Trade in Bananas: Obstacles and Opportunities for Progressive Social Change in the Global Economy." *Agriculture and Human Values* 17:65–73.
Murray, Jill. 1998. *Corporate Codes of Conduct and Labour Standards*. ACTRAV Working Paper Geneva: ILO.
Myers, Norman. 2004. *Banana Wars*. London: Zed Press.
Naranjo, Ana Victoria. 1999. "Health Risks for Women Banana Workers." In Foro Emaus 1999b:11–14.
Narayan, Deepa. 1999. "Bonds and Bridges—Social Capital and Poverty." Poverty Research Working Paper No. 2167, World Bank, Washington, D.C.
Nash, June. 2000. "Postscript: To Market To Market." In *Artisans and Cooperatives: Developing Alternate Trade for the Global Economy*, ed. by Kimberly Grimes and B. Lynne Milgram, 175–79. Tucson: University of Arizona Press.
Neuman, W. Lawrence. 2007. *Basics of Social Research*. 2nd ed. Boston: Pearson.
Nichols, Alex, and Charlotte Opal. 2005. *Fair Trade: Market-Driven Ethical Consumption*. Palo Alto, CA: Sage.

NREA (National Economic Research Associates). 2003. *Banana Exports from the Caribbean Since 1992.* Washington D.C.: NREA.

O'Connor, Martin, ed. 1994. *Is Capitalism Sustainable?* New York: Guilford Press.

OECD. 2006. *Investment Policy Reviews: Caribbean Rim (Antigua, Barbuda, Grenada, and St. Lucia).* Paris: OECD.

Offe, Claus. 1985. "New Social Movements: Challenging the Boundaries of Institutional Politics." *Social Research* 52, no. 4:817–68.

OkéUSA 2007. www.okeusa.com.

Oxfam. 2004. "Cool Planet." Children's Web page on Oxfam site, http://www.oxfam.org .uk/coolplanet/kidsweb/banana/

Palencia, Tania. 1997. *Creación de una instancia coordinadora para detener el impacto de la expansion bananera incontrolada.* San José, Costa Rica: Fundación Arias para la Paz y el Progreso Humano.

Parker, Liz, and James Harrison. 2005. "Brief Summary of 'Bananas: Differentiating Tariffs According to Social, Environmental and/or Economic Criteria.'" *IBCII Preparatory Papers,* 118–21.

Pearson, Neale J. 1987. "Honduras." In Greenfield and Maram, 463–94.

Peckenham, Nancy, and Anne Street, eds. 1985. *Portrait of a Captive Nation.* New York: Praeger.

Pedraja Toman, René. 1987. "Colombia." In *Latin American Labor Organizations,* ed. by G. Michael Greenfield and Sheldon Maram, 179–212. Westport, CT: Greenwood Press.

Perez, Veronica (FLO). 2008. Communication with author, January 21.

Perillo, Robert. 1998. "Long-Running Dispute Over the Right to Organize." *Report on Guatemala* 19, no. 3:6–9.

———. 2000. "The Current Crisis in the Latin American Banana Industry." In *A Strategic Analysis of the Central American Banana Industry: An Industry in Crisis,* ed. by Robert Perillo and Trejos, 11–20.

———. 2005. USLEAP presentation. North American Preparatory for International Banana Conference II, Washington, D.C., February 2.

Perillo, Robert, and Maria Eugenia Trejos, eds. 2000. *A Strategic Analysis of the Central American Banana Industry: An Industry in Crisis.* Chicago: USLEAP for COLSIBA and AFL-CIO/ACILS.

Pfeiffer, Rudi (BanaFair). 2008. Communication with USLEAP, January 30.

Phillipps, Sharon. 1987. "Panama." In Greenfield and Maram, 577–96.

Polaski, Sandra. 2003. *How to Build a Better Trade Pact with Central America.* Carnegie Endowment Issue Brief.

Polanyi, Karl. 2001. *The Great Transformation: The Political and Economic Origins of Our Time.* Boston: Beacon Press.

Polletta, Francesca. 1999. "Free Spaces in Collective Action." *Theory and Society* 28, no. 1:1–38.

Polletta, Francesca, and James M. Jasper. 2001. "Collective Identity and Social Movements." *Annual Review of Sociology* 27:283–305.

Potobsky, Geraldo von. 1998. "Freedom of Association: The Impact of Convention No. 87 and ILO Action." *International Labour Review* 137, no. 2:195–222.

Praxiom Research Group. 2007. "Comparison of ISO 14001 2004 and 14001 1996." http://www.praxiom.com/iso-14001-new.htm.

Purcell, Trevor W. 1993. *Banana Fallout: Class, Color and Culture Among West Indians in Costa Rica.* Los Angeles: Center for Afro-American Studies, University of California.

Ransom, David. 1999. "Fruit of the Future." *New Internationalist*, no. 317 (October).

———. 2001. *The No Nonsense Guide to Fair Trade*. Oxford, U.K.: New Internationalist/Verso.

Raynolds, Laura T. 2000. "Re-embedding Global Agriculture: The International Organic and Fair Trade Movements." *Agriculture and Human Values* 17:297–309.

———. 2003. "The Global Banana Trade." In Striffler and Moberg 2003:23–47.

———. 2007. "Fair Trade Bananas: Broadening the Movement and Market in the U.S." Chapter 5 in Raynolds, Murray, and Wilkinson 2007.

———. 2008. "The Organic Agro-Export Boom in the Dominican Republic: Maintaining Tradition or Fostering Transformation?" *Latin American Research Review* 43, no. 1:161–84.

Raynolds, Laura, and Douglas Murray. 1998. "Yes, We Have No Bananas: Re-Regulating Global and Regional Trade." *International Journal of Sociology of Agriculture and Food* 7:7–44.

Raynolds, Laura, Douglas Murray, and John Wilkinson. 2007. *Fair Trade*. New York: Routledge.

Real, Michael. 1996. *Exploring Media Culture*. Thousand Oaks, CA: Sage.

Reich, Robert. 2007. *Supercapitalism*. New York: Knopf.

Rogers, Tim. 2004. "Small Coffee Brewers Try to Redefine Fair Trade." *Christian Science Monitor*, April 13.

Rojas Valverde, Álvaro. 2003. "10 años de lucha del Foro Emaús." In *Foro Emaús* 2003: 35–45.

Romero, Simon. 2007. "Colombia May Extradite Chiquita Officials." *New York Times*, March 19.

Rose, Fred. 2000. *Coalitions Across the Class Divide*. Ithaca, NY: Cornell University Press.

Rostow, W. W. 1971. *The Stages of Economic Growth*, 2nd ed. Cambridge: Cambridge University Press.

Russo-Andrade, R., and C. E. Hernández. 1995. "The Environmental Impact of Banana Production Can Be Diminished by Proper Treatment of Wastes." *Journal of Sustainable Agriculture* 5, no. 3:5–13.

Sass, Robert. 2000. "Agricultural 'Killing Fields': The Poisoning of Costa Rican Banana Workers." *International Journal of Health Services* 30, no. 3:491–514.

Scher, Peter. (U.S. Special Trade negotiator). 1998. Press conference, November 11.

Schroyer, Trent, and Thomas Golodik. 2006. *Creating a Sustainable World*. New York: Apex Press.

Seidman, Gay. 2007. *Beyond the Boycott: Labor Rights, Human Rights and Transnational Activism*. East Palo Alto, CA: Russell Sage Foundation.

Shorrock, Tim. 2003. "Labor's Cold War." *The Nation*, May 19:15–22.
Shreck, Aimee. 2002. "Just Bananas? Fair Trade Banana Production in the Dominican Republic." *International Journal of Sociology of Agriculture and Food* 10, no. 2:13–23.
———. 2004. "International Standards, Social Certification, and Sustainable Agriculture." Presentation at the Conference on Cultivating a Sustainable Agricultural Workplace Conference, Troutdale, Oregon, September 13.
———. 2005. "Resistance, Redistribution and Power in the Fair Trade Banana Initiative." *Agriculture and Human Values* 22:17–29.
Sikkink, Katherine. 2004. *Mixed Messages: U.S. Human Rights Policy and Latin America*. Ithaca, NY: Cornell University Press.
Slocum, Karla. 1996. "Producing Under a Globalizing Economy: The Intersection of Flexible Production and Local Autonomy in the Work, Lives, and Actions of St. Lucian Banana Growers." Ph.D. diss., University of Florida.
———. 2003. "Discourses and Counterdiscourses on Globalization and the St. Lucian Banana Industry." In Striffler and Moberg 2003:253–85.
Smith, Jackie, and Dawn Wiest. 2005. "The Uneven Geography of Global Civic Society: National and Global Influences on Transnational Association." *Social Forces* 84, no. 2:621–52.
Smith, Jackie, and Hank Johnston, eds. 2002. *Globalization and Resistance*. Lanham, MD: Rowman & Littlefield.
Smith, Katie. 1994. *The Human Farm: A Tale of Changing Lives and Changing Lands*. West Hartford, CT: Kumarian Press.
Snow, David, and Robert Benford. 1988. "Ideology, Frame Resonance, and Participant Mobilization." *International Social Movement Research* 1:197–217.
———. 1992. "Master Frames and Cycles of Protest." In *Frontiers in Social Movement Theory*, ed. by Aldon Morris and Carol McClurg Mueller, 133–55. New Haven: Yale University Press.
Snow, David A., and Doug McAdam. 2000. "Identity Work Processes in the Context of Social Movements: Clarifying the Identity/Movemnent Nexus." In *Self, Identity and Social Movements*, ed. by Sheldon Stryker, Timothy J. Owens, and Robert W. White, 41–67. Minneapolis: University of Minnesota Press.
Snow, David, E. Burch Rochford Jr., Steven K. Worden, and Robert Benford. 1986. "Framework Alignment Processes, Micromobilization and Movement Participation." *American Sociological Review* 51:464–81.
Soluri, John. 2005. *Banana Cultures*. Austin: University of Texas Press.
Spalding, Hobart. 1977. *Organized Labor in Latin America*. New York: New York University Press.
Spaulding, Alan. 2005. "The Retail/Supermarket Chain: The U.S. Worker Perspective." IBCII Preparatory Conference, Washington, D.C., February 2.
Staggenborg, Suzanne. 1989. "Stability and Innovation in the Women's Movement: A Comparison of Two Movement Organizations." *Social Problems* 36, no. 1:75–92.
Stephen, Craig. 2005. "Race to the Bottom: Chiquita's Perspective." IBCII North American Preparatory Conference, Washington, D.C., February 2.

Stewart, Watt. 1964. *Keith and Costa Rica*. Albuquerque: University of New Mexico Press.

Stover, R. H., and N. W. Simmonds. 1987. *Bananas*. 3rd ed. Commonwealth Agricultural Bureau. London: Longman.

Striffler, Steve. 2003. "The Logic of the Enclave: United Fruit, Popular Struggle, and Capitalist Transformation in Ecuador." In Striffler and Moberg 2003:171–90.

Striffler, Steve, and Mark Moberg, eds. 2003. *Banana Wars*. Durham, NC: Duke University Press.

Sutton, Paul. 1997. "The Banana Regime of the European Union, the Caribbean and Latin America." *Journal of Interamerican Studies and World Affairs* 39, no. 2:5–36.

Swepston, Lee. 1998. "Human Rights Law and Freedom of Association: Development Through ILO Supervision." *International Labour Review* 137, no. 2:169–94.

T and G Publications. 2004. "Free Trade Farmers Fear Multinational Onslaught." *EPIU Bulletin* (U.K.), October 1.

Tarrow, Sydney. 1989. *Democracy and Disorder*. Oxford: Oxford University Press.

———. 1998. *Power in Movement: Social Movements, Collective Action and Politics*. Cambridge: Cambridge University Press.

Taylor, J. Gary, and Patricia J. Scharlin. 2004. *Smart Alliance*. New Haven: Yale University Press.

Thamotheram, Raj, ed. 2000. *Visions of Ethical Sourcing*. London: Financial Times and Prentice Hall.

Thiele, Stefen (Consultant, Foro Emaús). 2000. *La Anunciada Crisis Bananera y Sus Impactos en Costa Rica*. San José, Costa Rica: Foro Emaús.

Thomas-Slayter, Barbara. 2003. *Southern Exposure: International Development and the Global South in the Twenty-First Century*. West Hartford, CT: Kumarian Press.

Thompson, E. P. 1963. *The Making of the English Working Class*. Berkeley and Los Angeles: University of California Press.

Thompson, Ginger, and Nazila Fathi. 2005. "For Honduras and Iran, World's Aid Evaporated." *New York Times*, January 11:9.

Thompson, Robert. 1987. *Green Gold, Bananas and Dependency in the Eastern Caribbean*. London: Latin America Bureau.

Tilly, Charles. 1978. *From Mobilization to Revelation*. Reading, MA: Addison-Wesley.

Tiney, Juan (CLAC). 2007. "The Global Farmers Movement Against Neoliberal Globalization." Faculty Seminar on Globalization, Labor and Popular Struggles, Columbia University, New York, September 17.

Touma, Guillermo (FENACLE). 2005. Commentary. North American Preparatory for International Banana Conference II, Washington, D.C., February 2.

Touraine, Alan. 1981. *The Voice and The Eye*. Cambridge: Cambridge University Press.

———. 1985. "Introduction to the Study of Social Movements." *Social Research* 52, no. 4 (Winter):749–87.

TransFair. 2005a. "About Fair Trade." http://www.transfairusa.org.

———. 2005b. "Fair Trade Banana Cooperative Profile: Cerro Azul Ecuador." Pdf file available from author.

———. 2006. Web. http://www.transfairusa.org. Answer 15, "Frequently Asked Questions." (Accessed February 15, 2008.)
TransFair/COLSIBA. 2005. "Acuerdo de colaboración estratégica, COLSIBA y Trans-Fair." http://www.colsiba.org/node/29. Junio 8.
Travers, Max. 2001. *Qualitative Research Through Case Studies*. London: Sage Publications.
Trejos, Maria Eugenia. 1996. *La parcialidad de la calidad total: El caso de las bananeras en Honduras*. San José, Costa Rica: ASEPROLA.
Trouillot, Michel-Rolph. 1988. *Peasants and Capital: Dominica in the World Economy*. Baltimore, MD: Johns Hopkins University Press.
Tsing, Anna Lowenhaupt. 2004. *Friction: An Ethnography of Global Connection*. Princeton, NJ: Princeton University Press.
U.K. Food Group. 2007. *Partnership Under Pressure*. May 23.
UNCTAD (The United Nations Commission on Trade and Development). Var. *Commodity Price Bulletin* http://www.unctad.org/Templates/Page.asp?intItemID=1889&lang=1.
United States of America v. Chiquita Brands International, Inc., Defendant, March 13, 2007.
USLEAP (U.S. Labor Education in the Americas Project). 1999. "Oldest Union in Guatemala Under Attack by Del Monte," USLEAP 3 and 4 (December):1.
———. 2000. "Del Monte Workers Return to Work but Victory Is Bittersweet," USLEAP 3 (December):1, 6.
———. 2001a. *"First Come, First Served": A Race to the Bottom for Banana Workers*. Background memo. Chicago, March 13.
———. 2001b. "Chiquita Signs 'Historic' Worker Rights Agreement with Unions and the IUF." USLEAP 2 (August):1, 6–7.
———. 2002a. "Ecuador Banana Workers Strike in Critical Fight Against Race to the Bottom." USLEAP (April) 1:1–2.
———. 2002b. "Banana Workers Fight Ends for Now; New Campaign Begins." USLEAP (December) 3:1,3.
———. 2005. "DOLE: One Conflict Resolved; New One Erupts." USLEAP (October) 3:8.
———. 2006a. "Shifting Sands in the Banana Sector." USLEAP (March) 1:1–8.
———. 2006b. "Buenos amigos." USLEAP (March) 1:1.
———. 2007. "Banana Unions Launch Initiative Against Chiquita." USLEAP (June) 2:3.
USTR. 1995. *Kantor Will Use WTO Mechanisms*. Press Release, September 27.
USTR. 1998a. "U.S. Officials Discuss EU Banana-Import Policies." Worldnet Dialogue transcript, U.S. Information Service, December 10.
———. 1998b. Press release, December 16.
Vado, Mario Blanco. 1998. *Centroamerica: Reformas al derecho colectivo de trabajo*. San José, Costa Rica: ASEPROLA.
Valelly, Richard M. 1989. *Radicalism in the States*. Chicago: University of Chicago Press.
Valticos, Nicholas. 1998. "International Labour Standards and Human Rights: Approaching the Year 2000." *International Labour Review* 137, no. 2:135–47.
van de Kasteele, Adelien. 1998. "The Banana Chain: The Macro-Economics of the Banana Trade." Paper delivered at the International Banana Conference, Brussels, Belgium, May 4–6.

van de Kasteele, Adelien, and Myriam van der Stichele. 2005. "Update on the Banana Chain." *IBCII Preparatory Papers*, 8–35.
Van Loo, Rory. 2004. "Fair Trade Bananas and Other Products Now Spreading in U.S." *Christian Science Monitor*, January 6.
VanderHoff Boersma, Francisco (UCIRI Councilor). 2005. *Antigonish* (June 20).
Vargas, Gerardo. 1998. "The Socio-Environmental Problems of Banana Plantations in Costa Rica." In Foro Emaús 1998d:4–7.
Vargas, Gerardo, and Hernan Hermosilla, eds. *Foro 2000: The Legacy of More than a Century of Banana Production.* 2nd ed. Foro Magazine. San José, Costa Rica: Foro Emaús.
Vogel, David. 1978. *Lobbying the Corporation.* New York: Basic Books.
Wallerstein, Immanuel. 1976. *The Modern World-System.* New York: Academic Press.
Waridell, Laure. 2002. *Coffee with Pleasure: Just Java and World Trade.* Montreal: Black Rose Books.
Waridell, Laure, and Sara Teitelbaum. 1999. *Fair Trade in Europe: Contributing to Equitable Commerce in Holland, Belgium, Switzerland, and France.* Montréal: Equiterre.
Warning, Matthew. 2006. Commentary in *Buyer Be Fair: The Promise of Fair Trade Certification.* DVD. John DeGraff, producer, Fox Wilmar Productions. Distributed by Bullfrog Films.
Warning, Matthew, N. Key, and W. Soo Hoo. 2002. "Small Farmer Participation in Contract Farming." Presented at the Western Economics Association International Annual Meeting, Seattle, Washington, July 1.
Wedin, Ake. 1986. *International Union Solidarity and Its Victims: The Costa Rican Case.* Research Paper #43, Stockholm: Institute of Latin American Studies.
Wellman, Barry, ed. 1999. *Networks in the Global Village.* Boulder,CO: Westview Press.
Welsh, Barbara. 1996. *Survival by Association: Supply Management Landscapes of the Eastern Caribbean.* Montreal: McGill-Queen's University Press.
Wesseling, Catharina. 1997. *Health Effects from Pesticide Use in Costa Rica: An Epidemiological Approach.* Stockholm: Karolinska Medico Institutet.
Wesseling, Catharina, Anders Ahlbom, D. Antich, A. C. Rodriguez, and R. Castro. 1996. "Cancer in Banana Plantation Workers in Costa Rica." *International Journal of Epidemiology* 25:1125–31.
Wesseling, Catharina, Christer Hogstedt, Jai-Dong Moon, Rob McConnell, Matthew Keifer, Linda Rosenstock, and Anders Ahlbom. 2002. *Trabajadores bananeros en Costa Rica y el riesgo del cáncer: Una actualización de cinco años.* Heredia, Costa Rica: Instituto Regional de Estudios en Sustancias Tóxicas, Univ. Nacional.
Wheat, Andrew. 1996. "Toxic Bananas." *Multinational Monitor* (September):9–15.
Whelen, Tensie. 2003. "Holding Chiquita Accountable for Environmental Excellence." *Rainforest Alliance Verification Statements* (July).
Wickham-Crowley, Timothy P., and Susan Eckstein. 2003. *What Justice? Whose Justice?* Berkeley and Los Angeles: University of California Press.
Williams, Oliver F., ed. 2000. *Global Codes of Conduct: An Idea Whose Time Has Come.* Notre Dame, IN: University of Notre Dame Press.

Williams, Peter. 2007. "Preliminary Findings on FLO-union Relations." Circulated paper, Fair Trade Foundation, London, October.
Willmott, Michael. 2001. *Citizen Brands: Putting Society at the Heart of Your Business.* New York: John Wiley and Sons.
WINFA. 2003. "Fair Trade in the Windward Islands." Mimeo, St. Vincent: WINFA.
Wirada, Howard, and Harvey F. Kline. 1985. *Latin American Politics and Development.* Boulder, CO: Westview.
Wisniewski, Linda C. 2006. "Female Banana Packers Gain Grounding in Rights." *WeNews,* November 23.
Wolf, Eric R. 1968. *Peasant Wars of the Twentieth Century.* New York: Harper and Row.
Wood, Leslie. 2004. "Bridging the Chasms: The Case of Peoples' Global Action." In Bandy and Smith, 95–120.
Worobetz, Kendra. 2000. "The Growth of the Banana Industry in Costa Rica and Its Effect On Biodiversity." San José: Foro Emaus. http://www.foroemaus.org/english/history/02_02.html.
Yanz, Linda, and Bob Jeffcott. 2001. "Bringing Codes Down to Earth." *International Union Rights* 8, no. 3.
Zald, Mayer N., and John D. McCarthy. 1979. *The Dynamics of Social Movements: Resource Mobilization, Social Control and Tactics.* New York: Winthrop Publishers.
Zalla, Jeffrey (Corporate Responsibility Officer, Chiquita Brands). 2000. Presentation at SAI Annual Conference, New York, December 7.
Zepeda, German. 2005. "Trade Union Perspective, Latin America." North American Preparatory for International Banana Conference II, Washington, D.C., February 2.
———. 2007. Zepeda to Ron Oswald, Secretary General, IUF. Junio.
Zonneveld, Luuc. 2007. "Fairtrade Labeling—FTL Background and Perspectives on Hired Labour." Presentation at "Working Together for Fair Labour Conditions," FLO and trade union workshop, Bonn, Germany, June 25.

Interviews

Arana Toscana, Ing. Mario (former Dole Production Manager). 2001. Interview with author, July 19.
Arguedas Mora, Carlos (Secretaría General, SUTAP). 1999b. Interview with author, July 2.
Barrantes Cascante, Ramón (Secretaría General, SITAGH). 1999a. Interview with author, August 6.
———. 2004. Interview with author, June 29.
Barrientos, Irene (International Relations, UNSITRAGUA). 2001. Interview with author, July 11.
Bermúdez Umaña, Gilberth (Coordinator, COLSIBA). 1999. Interview with author, August 5.
———. 2004. Interview with author, July 5.
Bowman, Anton (WINFA). 2005. Interview with author, February 2.

Fieldman, Bruce. 2002. Interview with author, April 5.
Garcia Miranda, René. 2004. Interview with author, July 3.
Guerra, Jiménez (Chiquita worker). 2004. Interview with author, July 3.
Hermosilla, Hernán (Vice Director, Foro Emaús). 1999. Interview with author, August 5.
——. 2004. Interview with author, June 27.
Holst, Eric (Rainforest Alliance). 1999. Interview with author.
Kearney, Neil (Secretary General, ITGLWF). 2002. Interview with author, December 7.
Lucio, Diogenes (FENOCIN/CLAC). 2007. Comments to author, September 18.
Mancilla, Carlos (Secretaría General, CUSG). 2001. Interview with author, July 16.
Martínez, Mel (Secretaría General, SITRABI). 1989. Interview with author, August 18.
McCracken, Carrie. 2004. Interview with author, June 30.
Morales, David (Secretaría General, FESTRAS). 2001. Interview with author, July 9.
Oswald, Ronald (Secretaría General, IUF). 2006. Comments to author, February 10.
Quezada, Victor (Central American labor sociologist). 2004. Interview with author, June 30.
Ramírez, Hugo (Chiquita Brands). 1999. Interview with author, August 13.
Rojas Valverde, José Manuel. 2004. Interview with author, June 28.
Roldan, Carmen (Rainforest Alliance). 1999. Interview with author, August 14.
Rosenthal, Jonathan. 2007. Interview with author, June 16.
Thiele, Stefen (Consultant, Foro Emaús). 1999. Interview with author, August 13.
Traub-Werner, Marion (STITCH). 2001. Interview with author, July 11.
Trejos, Maria Eugenio. 2004. Interview with author, June 28.
Valenciano, Enrique (Limón Social Pastorate). 1999. Interview with author, August 10.
Vargas Castañeda, Auria (Executive Committee, SITRACHIRI). 2004. Interview with author, July 2.
Vázquez, Enrique (Vice President of Legal and Political Affairs, Chiquita CR). 2004. Interview with author, July 5.
Wille, Chris (Regional Director, Rainforest Alliance). 1999. Interview with author, August 6.
——. 2004. Phone interview with author, July 1.
Workers of Sarapiquí. 1999. Various interviews with author, August 11. Available on request.
——. 2004. Various interviews with author, July 4. Available on request.
Workers of Siquirres. 1999. Various interviews with author, August 9. Available on request.
Workers of Sixaola. 2004. Various interviews with author, July 3. Available on request.
Wunderlich, Christopher (Rainforest Alliance). 1999. Interview with author, August 25.
Zepeda, German. 2006. Interview with author, March 17.
Zúñega, Martin (CORBANA Director). 2004. Phone interview with author, June 30.

Index

3F (Denmark), 52, 54, 152, 195, 220
ACP countries, 5, 17, 85–86, 89, 91, 197; bananas, 28, 38, 85–86, 88, 124, 129, 164; trade policy, 89, 92–93, 197–204
AFL-CIO, 54, 112, 213
AgroFair, 38–39, 41, 46, 52, 55, 59, 113, 155, 182, 184–85, 217, 228n19, 237n25, 241n13
AIFLD, 54, 101, 228–29n9, 234n3
Arguedas Mora, Carlos, 108, 217

BanaFair, 38–39, 43, 55, 62, 152, 182, 194, 213–14, 217, 220, 227n7, 239n4
banana alliance, 11–13, 47, 59–68, 92, 97, 117, 136, 156, 173, 187, 196, 205–22; requirements, 66–67, 214; trade policy, 204
Banana Link, 3, 38, 53, 62, 145, 152, 156, 182, 190, 213, 217, 220, 228n8, 234n8, 238n12, 240n19
bananas: consumption, 1, 4, 45, 84; environmental impact, 70–83; expansion, 72, 98, 223n1; importers, 42–43, 84; price, 3, 7, 26–28, 36, 42, 75, 84, 95, 119, 129–31, 158, 180, 184, 193, 208, 213, 217, 220, 223n2, 227n11, 234n24, 238n8; varieties, 72–73. *See also* fair trade; trade
Barrantes, Ramón, 9–10, 53, 108, 110, 138, 142, 230n7
Barratt Brown, Michael, 34, 50, 215
Barrientos, Irene, 167, 169
Belize, 53, 115, 119–20, 124, 166, 198–99, 201, 242n6; Fyffes, 120
Bermúdez, Gilberth, 77, 102, 106, 113, 139, 142, 153, 198, 233n12, 235n11, 239n17
Bonita, 24, 26, 54, 190, 241n7. *See also* Noboa

Bowman, Anton, 6, 93, 121, 123, 136, 208, 235n3
Brazil, 23, 166, 189, 199, 209, 226n33, 227n10, 234; Del Monte, 226n34; exports, 4
Britain. *See* United Kingdom
Bureau Veritas Certification (BVQT) 157, 159

Cameroon, 4, 23, 124, 179; Del Monte, 166, 234n25; Dole, 91, 189, 220
Campaign for Labor Rights, 52, 56, 243n6
Canary Islands, 1, 85, 198
Caribana, 27–28, 141, 151–53, 226n31, 238n7, 240n10
Caribbean bananas, 117–36; Certified Growers Programme, 125; Fair Trade, 126ff
CARIFORUM, 202
CEPIBO, 184, 193–94
CGT (France), 52, 63, 194–95, 220
Chambron, Anna, 48, 156, 159–66, 172, 179–80, 230n13, 238n12, 239n6
Change to Win, 213
child labor, 159, 168, 175. *See also* Human Rights Watch
Chiquita, 19–24, 124, 133; bankruptcy, 138; Colombia, 106–8, 139; COLSIBA Accord, 137–56; corporate responsibility, 156; Costa Rica, 106–8, 139; Ecuador, 6, 115; environment, 109, 151–52; Fair Trade, 39–40, 82, 142, 185–86, 218–19; Guatemala, 49–50, 104, 112, 142, 166; Honduras, 61, 109, 178; independent producers, 27, 59, 143–44; labor

Chiquita (cont.)
relations, 110, 115–16, 143, 147–53; Panama, 104, 147; Rainforest Alliance, 39–40, 75–77, 148, 157–63; sales, 24, 87; subcontracting, 44; trade policy, 40, 86–93, 198, 200, 202. See also United Fruit Co.
CLAC, 46, 51–52, 65, 208, 220, 240n17; Fair Trade, 53, 129, 185; Trans-Fair, 186; union relations, 53, 186, 216
Clinton, Pres. William, 90, 233n16
Coats, Stephen, 54–55, 170, 243n4
Colombia, 18, 102, 165; AUC, 21–22, 105; exports, 4, 200; Chiquita, 19–22, 88, 91, 142–44, 162; coops, 180, 183; labor relations, 93–99, 105–10, 150, 176; trade policy, 202
COLSIBA, 52–53, 21, 105–7, 214, 216; Chiquita, 115–16, 138, 152–53, 168; Chiquita Accord, 44, 137ff; Colombia, 108; coops, 181; Costa Rica, 153; Dole, 186, 189, 194–96; Ecuador, 183, 190; environment, 109, 152, 216; Fair Trade/FLO, 65, 115, 185–86, 215, 220; Honduras, 148; networking, 61, 154–55, 196, 210–11; Peru, 194–96; product certification, 168; small farmers, 136; trade policy, 92, 173, 197–200, 203, 207, 209; women, 150
consumers, 43–44, 217ff
Coop America, 33, 56, 186
cooperatives. See fair trade
CORBANA, 77, 93, 164–65
corporate codes. See product certification; see also under specific companies
corporations. See transnational banana corporations (TNCs)
Costa Rica, 15, 19, 26–28, 109, 199; Chiquita, 39–40, 139, 142–53; Del Monte, 110, 115, 145; Dole, 91, 110, 191–97; environment, 73–78, 82, 152, 158, 162; exports, 4; Fair Trade, 10, 180–82; labor relations, 53, 98–99, 145, 172–74; network, 62, 216–17; promotion, 163–64; trade, 88. See also Solidarista associations
Cotonou Accord, 92, 197–98
CTM Altromercato (Italy), 52, 56, 181, 220, 228n18, 240n18
Cuperlier, Sylvain, 40, 163, 170, 193–96
CUSG (Guatemala), 112, 167
CUT (Colombia), 108

DBCP, 73–75, 108, 192, 230n7
Del Monte, 9, 19, 23–27, 93, 163, 198; environment, 40, 75, 217; labor relations, 99–100, 105–6, 110–12, 145, 153, 166; sales, 24, 87
development theories, 11, 48ff
Divney, Thomas, 157–58; 227n13
Dole, 15, 18, 23, 26–28, 87, 189–97; environment, 39, 74–78, 162, 192ff; Fair Trade, 82, 183–86, 193ff; independent producers, 144; labor relations, 102, 104, 109–10, 115, 153, 166, 191ff; sales, 24, 87; trade policy, 91–93, 198
Dominica, 17, 121–26; exports, 119; Fair Trade, 43, 129–32, 200; trade, 200
Dominican Republic, 15, 80; environmental practices, 80, 133–34; exports, 119; Fair Trade, 84, 95, 133, 180, 183; trade, 124, 199

Ecuador, 25–26, 143, 200, 225n28; Chiquita, 144; coops, 182–83; Dole, 91, 166, 189–96; environment, 80, 133–34; exports, 4, 223n1; Fair Trade, 42, 180, 182–85, 234n23; labor relations, 109–10, 138, 145, 149, 189–91, 194, 211, 241n4; promotion, 163–65; small farmers, 6–7; trade, 89, 92, 201–2, 233n18
EFTA, 35, 56, 226n4
environment, 35, 39–40, 70–82, 207; Caribbean practices, 123, 128; Chiquita, 20, 75, 151–52, 157–58; Dole, 192ff; Fair Trade, 5–7, 40–43, 78–81, 129–30, 215; independent producers, 26, 165; network, 62,

216; pesticides, 72–76; union policy, 108–9, 151–52, 212, 217. *See also* Fair Trade/labeling (FLO), environment; transnational banana corporations (TNCs), environment
EPAs, 199, 201–4
Equal Exchange, 11, 32–34, 46, 52, 55, 186
ETI, 152, 157–60, 163, 216, 219
EU (European) Commission, 93, 199–201
Eurepgap, 158, 161–63, 183, 234n24
EUROBAN, 38–40, 52–53, 115, 156, 207, 228n8; Chiquita, 39–40, 75, 138, 153; Del Monte, 110; Dole, 194–96; Ecuador, 190; Fair Trade/FLO, 38, 43, 65–66, 220; membership, 228n8; network, 61, 210; trade policy, 92, 197–205
EU trade regime, 86–96, 124, 157, 197–202

fair trade, 1, 5–6, 10–12, 30–33, 44, 46, 55, 62, 66–67, 82, 203–4, 210, 214–15, 225n30, 240n17
Fair Trade Federation (U.S.), 56, 186
Fair Trade Foundation/FTF (U.K.), 32, 43, 55, 126, 132, 182, 187, 200, 215, 240n19
Fair Trade/labeling (FLO), 30–47, 50, 158, 164, 190, 206–12, 239n1, 239n4, 239n6; 240n10, 240n13, 240n15, 240n18, 243n8; banana standards, 40–43; Caribbean, 126ff, 200–201; consumers, 43–44; cooperatives, 179–84; critique, 33–35, 237n22; Dole, 189, 193–94; Economic Development, 126; environment, 78–81, 231n14; label, 11, 13, 32, 37–38, 207; network, 55–56, 62, 213; plantations, 185ff, 218–19; price, 43, 84, 227n8, 227n11, 227n17; sales, 45, 59, 180; size, 33, 52–53; social development, 113–15; trade, 92–96, 204; transnational banana corporations, 68, 240n17; unions, 44–46, 65–67, 113, 135–36, 145, 185–88, 215, 220
FAO, 158, 160, 165, 179, 203, 205, 232n5, 237n22
FENACLE, 189–91, 194–96

Finca 6 (Dominican Republic), 80–81, 95, 133–35, 180, 234n26
FLO-CERT, 35, 41, 114–15, 178, 183, 187–88, 215, 238n1
FOB, 18, 227n8; defined, 234n26
Fordist model, 102, 234n4
Foro Emaús, 39, 56, 62, 74–77, 216–17, 227n12, 230n7
France, 1, 17, 24, 28, 32, 45, 52, 54, 85, 89, 119, 174, 194, 218, 220, 224n8, 228n8, 232n2
free association, 66, 152–54, 166–67, 173–77; Fair Trade, 178–188; 215
Fyffes, 14, 17–18, 19, 24, 27–29, 86–87, 90–92, 115, 120–21, 125–26, 160, 166, 198, 223n1, 224n6, 228n19

GATT, 87–90, 232n11, 233n14
GAWU (Ghana), 113–15, 127, 184, 201, 203, 212
Gearhart, Judy, 159–60, 171, 177
Geest, 14, 17, 27–29, 86, 121–26, 208, 223n1, 224n7, 236n6
Germany, 32, 35, 38, 43, 44, 164, 187, 194, 213, 218, 224n8, 226n32, 228n18, 232n2
Global Exchange, 56, 92, 220
global framework accord. *See* Chiquita; COLSIBA; IUF
Greece, 1
Grenada, 51, 121–22
Gros Michel banana variety, 71, 73
Grossman, Lawrence, 25, 80, 86, 120–23
Guatemala, 26; Chiquita, 15–19, 104, 144, 147, 153; Dole, 198–99; exports, 4, 223n1; independent producers, 225n26; labor relations, 103–6, 110–12, 154, 230n12

Harari, Raul, 72, 76–77, 102, 146, 149, 154, 169, 230n5, 230n11, 234n4, 238n11, 238n16
Havelaar, 31–32, 35, 38, 43, 52, 65, 80, 226n2, 226n4, 227n7, 228n8

hired labor. *See* Fair Trade/labeling (FLO), cooperatives
Honduras: Chiquita, 15, 18–19, 61, 82, 109, 143, 153–54, 186, 238n7; exports, 4; labor relations, 63, 100–2, 105–6, 150, 230n12; trade, 89, 233n18
Human Rights Watch, 145; child labor, 190–91, 241n3
Hurricane Mitch, 26–27, 89, 109–13, 209

identity, 12, 57–59, 63–69, 79–82, 97–98, 108–9, 121, 124, 130, 135, 154–56, 163, 182, 196, 206–7, 214–16, 220–21, 229n2–30n15, 237n25
IFAT, 35, 56, 219
IFOAM, 158, 160–61, 219, 237n23
ILO: principles, 42, 153, 157–59, 162, 165–68, 173–74, 186, 188, 209; conventions, 113–15, 153, 166, 175–79, 204, 207, 234n2, 235n10, 239n2, 240n12
import licenses, 5, 27, 86–87, 90–95, 124, 227n11, 232n2, 232n4, 232n10, 233n14, 236n9
independent producers, 5, 8–9, 14, 22, 25–29, 32, 39, 82, 91, 100, 104, 113, 138, 142–44, 148, 158, 163–68, 191, 206, 209, 225n26, 225n30, 238n8
Institute for Agriculture and Trade Policy, 32
Integrated Crop Management, 76–79, 158
International Banana Conference, 115, 207–10
ISEAL, 239n6
ISO 14001, 75, 157–59, 162, 230n10
ITGLWF, 170, 172, 243n4
IUF, 52, 54, 60–61, 65, 108, 110, 113–16, 148, 153, 155, 177, 187–89, 195, 210–17, 228n8, 229n6, 229n9, 237n1, 243n4
Ivory Coast, 4, 91, 124, 195, 199, 201

Jamaica, 15, 17–18, 118–20, 198–99, 201
John XXIII Center (Costa Rica), 102, 106

joint bodies/committees, 66, 115, 176, 183, 187

labor codes, 100–102, 169, 174
Lamb, Harriet, 38, 43, 74, 105, 126, 132, 179, 182
Latin America Working Group, 213, 243n6
licenses. *See* import licenses
Lomé Convention, 85, 92, 123–24, 232n1
Longley, Susan, 149, 177–80, 212

Maquila Solidarity Network, 171–72, 239n6, 243n6
marijuana cultivation, 125, 237n20
McCracken, Carrie, 78, 145, 149, 181, 231n16, 238n9, 240n10
methodology, 67
Mexico, 18, 30–31, 89, 177, 213–14, 227n10; exports, 119
Moberg, Mark, 123–24, 127–29, 236n18
Munguía, Iris, 74, 106, 149–51, 230n6

national producers. *See* independent producers
nematicides, 73, 122, 146, 165, 230n5
neoliberal system, 13, 34, 118, 197
network. *See* social network
New International Economic Order, 31
New Social Movement theory, 57, 63
NGOs, 10–12, 52ff, 59, 145, 195, 197; Chiquita/Rainforest, 75, 163, 227n13; Fair Trade, 33, 43–44, 46, 62, 77, 82, 170, 226n3; unions, 55, 66, 150, 170, 212ff
Nicaragua, 26, 32, 88; Dole, 18, 74, 104, 109, 187, 192, 202; labor relations, 53, 101–4, 144, 153, 176, 238n16; solidarity bananas, 226n3
Nicaragua Network, 56, 243n6
NISGUA, 56
Noboa, 24, 26, 87, 190–91, 200, 241n5, 241n7
north-south divisions, 61, 113

Oké Bananas: Equal Exchange, 38, 46, 55, 59, 155, 182, 214, 220. *See also* Equal Exchange
opportunity structure. *See* structural opportunity
organic bananas, 32, 118, 158–61, 180, 219, 231n4, 235n3, 240n9, 240n10, 240n14, 241n11; Dole, 40, 193; Fair Trade, 80, 133–34, 182–84; production, 237nn21–23
Oswald, Ronald, 189, 237
Oxfam, 31–32, 35, 43, 46, 52, 55–56, 66, 130, 186, 201; unions, 66, 215–16

Panama, 3, 15, 202; exports, 2, 4, 232n12, 233n14, 233n20; labor relations, 26, 99–105, 147–48, 176, 230n12
Panama disease, 16, 72
Parker, Liz, 197, 203, 218, 242n8
peasant producers. *See* small farmers
permanent committees. *See* Solidarista associations
Peru: Fair Trade, 180, 184, 193, 228; labor relations, 44, 189, 194–96; organics, 32, 193, 241n11
pesticides, 2, 30, 39–40, 71–79, 90, 158; Caribbean use, 80, 123, 208; Fair Trade, 41–42, 79; unions, 73, 108–9, 216. *See also* DBCP; environment
Peuples Solidaires, 54, 195
Philippines, 4, 23, 158, 161, 193, 199, 223n1, 228n18, 231n19, 241n11
plastic bags, 72–81, 118, 123, 178, 217
Portugal, 1, 17, 85, 224n8
price. *See* bananas, price
product certification, 8, 34–35, 56, 75–79, 152, 157–72, 180, 183, 189, 218–19, 239n6; auditor role, 170; unions, 166–72, 230n12. *See also* Fair Trade/labeling (FLO)

Rainforest Alliance/SAN, 39–40, 77, 161–63, 219, 227; Chiquita, 75–77, 152, 157–59, 165, 171; social criteria, 115–16, 168, 231

Raynolds, Laura, 17, 34, 62, 66, 75, 86, 91, 125, 133–34, 180
Red Tomato, 46, 56, 184
Rice, Paul, 32, 50
Rosenthal, Jonathan, 11, 28, 32, 36, 46, 129, 162, 178, 182, 185–86, 216, 219, 226n4

SAI, 152, 159–61, 168, 170–72, 177; Dole, 189, 241n2
Sainsbury, 200–201
St. Lucia, 17, 51, 120–26, 236n11; exports, 119; Fair Trade, 127–31, 200
St. Vincent, 6, 17, 51, 80, 119–26, 235n1, 236n12, 236n15, 237n20; Fair Trade, 127, 130, 180
SGS, 157, 159, 170
Shreck, Aimee, 34, 80–84, 95, 134–35, 161, 179, 237n24, 240n8
sigatoka, 8, 25, 73, 76, 78–80, 99, 140, 158, 165, 241n11
SITRABI (Guatemala), 100, 103–4, 110–11, 153, 235n18, 238n13, 238n18
SITRAINAGRO (Colombia) 105–8
SITRATERCO (Honduras), 100–101, 109, 147, 154, 178, 238n15
small farmers, 2, 6–7, 14, 16–17, 29, 49–51, 67, 117, 178, 219, 240n15, 240n17; African, 220, 242n5; Caribbean, 5, 51, 80, 118–36; Fair Trade, 30–36, 40–42, 51–53, 59–60, 68, 96, 126–35, 180ff, 196, 207, 237nn24–25; identity, 65, 214; Latin America, 181–84; networked, 60, 74, 172, 206–11; organics, 133–35; trade, 82, 89–91, 197–204; unions, 135–36, 185–86, 209. *See also* COLSIBA
social movement, 57–69, 81–82, 206–22; requirements, 67–68; 214
social network, 12, 210–11, 216–17, 220; defined, 58, 60–63; farmers, 53, 82, 130, 184, 213–14; trade, 204–5; transnational activist network, 56–57, 62, 66, 68, 96, 152, 155, 171, 194–96, 213; union, 53–54, 105, 109, 115, 154, 190, 211–13

social premium, 3, 42, 127–36, 173, 182–83, 193, 209–10, 217, 240n13; unions, 46, 113–15, 173, 181–84
soil erosion, 2, 71, 122, 158
Solidarista associations, 25, 99, 102–3, 106, 110, 139, 152–53, 166, 176–77, 184, 191–95, 227n13, 234n8, 235n9
Solidarity Center, 52, 54, 212, 220
Spain, 1, 17, 45, 56, 85, 224n8, 228n8, 232n2
Standard Fruit, 15, 18, 29, 99–101. *See also* Dole
STITCH, 52, 55, 62, 150, 213, 220
structural opportunity, 12, 14, 57–59, 96, 172, 206–9, 220–21; environment, 70–71, 78–81; Fair Trade, 209–10; theory, 57–60, 67–88
subcontracting, 145–46, 194–95
supermarkets, 14, 26–29, 93, 161–62, 164, 195, 205–8, 243n9; Eurepgap code, 160–63; 234; Fair Trade, 44, 66, 132, 201; pricing, 3, 27, 236n18; product requirements, 123–29; 134–36, 208, 236n7, 236n13; workers, 28, 212, 240n19
Suriname, 44, 119–20, 124
Sustainable Agriculture Network. *See* Rainforest Alliance/SAN
SweatFree Communities, 56

Talamanca's Association of Small Producers (Costa Rica), 181
tariffs, 1, 38–40, 83–96, 109, 139, 143, 197–205, 224n8, 232n3, 232n5, 232n7, 242n2, 242n8; differential, 203
Teamsters, 52, 212
temporary workers. *See* subcontracting
third-party auditors. *See* product certification
trade, 83–96, 197–205; African supply management, 203; alliance proposals, 203; first come, first served, 90–92, 233n20; Geneva Agreement, 202; Latin bananas, 93, 198, 204, 224n8, 232n3, 232n12; top importers, 84; trade war, 87–92; U.S. approach, 84–94, 202. *See also* Fair Trade/labeling (FLO); *See also under individual countries*
trade unions, 2, 9–11, 25–26, 40, 51–54, 60, 66, 97–116, 137–55; Chiquita, 110, 115–16, 143, 147–53; Colombia, 93–99; corruption, 178, 186; Costa Rica, 53, 98–99, 145, 172–74; decision process, 209ff; Del Monte accords, 105–6, 110–12, 152; Dole accord, 195; Ecuador, 109–10, 138, 145, 149, 189–91, 194–96, 211, 241nn3–n7; environment, 108–12, 151–52, 212, 217; Fair Trade, 44, 46, 65–67, 113, 135–36, 145, 185–88, 215, 220; free association, 173ff; Foro Emaús, 216–17; Ghana, 113–15, 127, 184, 201, 203, 212, 227; Guatemala, 103–6, 110–12, 154, 230n12; history, 98–104; Honduras, 63, 100–102, 105–6, 150, 230; independent producers, 144–45; Ivory Coast, 201; national banana promotion, 164; network, 53–54, 105, 109, 115, 154, 190, 211–13; Panama, 26, 99–105, 147–48, 176, 230n12; Peru, 194–96; Philippines, 241n1; product certification: auditor evaluation, 169–72; replacing law, 168–69; replacing unions, 66, 115, 168, 176, 183, 187; small farmers, 135–36, 185–86; 209; trade policy, 92, 173, 197–200, 203, 207, 209; WINFA, 136, 200, 209–14; women, 63, 101, 106, 113, 116, 215. *See also under individual unions*; COLSIBA
TransFair, 32–38, 43, 50–52, 226n4, 239n7; CLAC, 46, 186; COLSIBA, 44, 73, 185–87, 243; Dole, 186, 195; social network, 62, 210, 215
transnational banana corporations (TNCs), 14–29, 148, 208, 223n11, 227, 242n3; CLAC critique, 240n17; environment, 39–40, 74–78, 117–18, 157–63; Fair Trade, 30, 68, 96, 187; labor relations, 104, 144–46, 166, 209; product certification, 157–61; sales, 87; trade,

86, 90, 124. *See also under individual companies*
Trejos, Maria Eugenia, 112, 145, 152

UCIRI (Mexico), 31–32, 226n1
UFCW, 52, 54, 212, 220, 243n4
unions. *See* trade unions
United Farm Workers (UFW), 189, 243n9
United Fruit Co., 14–24; environment, 72–73; labor relations, 98–105. *See also* Chiquita
United Kingdom, 17, 28, 43, 85, 89, 93, 121–23, 157, 164, 182, 200–201, 218, 233n17
U.N. Universal Declaration of Human Rights, 174
United States, 1, 100, 174–76, 223n1; trade, 84–93, 201–4, 212–13
U.S. Trade Representative, 176, 213, 239, 243; banana trade, 87–92, 233n19
United Students for Fair Trade, 56
UNSITRAGUA, 104, 154, 167, 196, 171; Alabama and Arizona struggle, 112
UROCAL (Ecuador), 182
USLEAP, 52, 55, 61, 66, 115, 152, 243nn6–7; Chiquita-COLSIBA accord, 138, 153, 238; Del Monte Campaign, 110ff; Dole, 191, 194–95, 241; Ecuador, 190; Fair Trade, 185; network, 138, 210–13

van de Kasteele, Adelien, 24, 27, 87, 223n1, 226nn32–34, 232n8, 237n21
VanderHoff, Franz, 31–33, 226n1
van der Stichele, Myriam. *See* van de Kasteele, Adelien
Vázquez, Enrique, 142–47, 162, 218, 226n31, 238n7, 238n10, 239n4, 243n8
Via Campesina, 51, 203, 211. *See also* CLAC

Volta River Estates (VREL, Ghana), 79, 95, 113–14, 127, 184, 220

Wal-Mart, 13, 27–28, 33, 161, 226n34
Whole Foods, 44, 82, 161–62, 239n3, 240n10
WIBDECO, 90, 125, 200–201
Wille, Chris, 39, 115, 163, 227n13
Windward Islands, 6, 17–18, 93, 118, 121–26; exports, 3, 119, 180; Fair Trade, 126–33; trade policy, 198–200
WINFA, 6, 51–52, 60, 132; Fair Trade issues, 129–30, 180, 200, 236n19; identity, 65, 121; networking, 210–11, 220; trade, 93, 197–205; unions, 136, 200, 209–14
Witness for Peace, 56, 243n6
women, 9, 28, 55, 149–51, 181, 213; health, 74–78, 97–98, 149, 231n23; production, 113, 124, 181, 236n8; unions, 63, 101, 106, 113, 116, 215; wage inequality, 183
worker benefits, 144, 148–49, 166, 209, 234–35; housing, 100, 103–8, 140, 181, 191; medical, 8, 114, 145–46
workers, 6–9, 97–116, 137–55, 179–84
working conditions, 70–71, 97–98, 146–48; child labor, 159, 168, 175, 190–91; health risks, 74–78, 97–98, 149, 231n17; subcontracting, 145–46; wages, 3, 9, 30, 104, 124, 139–54, 162, 166–68, 179–87, 208, 215, 227n13
World Bank, 49, 228n4
World Shops, 52, 55–56
WTO, 86–93, 114, 156, 197–203, 207, 232n11; dispute rulings, 89–90, 202

Zepeda, German, 44, 136, 142, 144, 147–48, 154, 180, 183, 185, 190, 198–99, 215, 237n1

About the Author

HENRY J. FRUNDT is a convener-emeritus of the Sociology faculty at Ramapo College, where he has been teaching for more than thirty-five years. Dr. Frundt's previous publications include the award-winning *Trade Conditions and Labor Rights: U.S. Initiatives, Dominican and Central American Responses*; *Refreshing Pauses: Coca-Cola and Human Rights in Guatemala*; *An Agribusiness Manual*; and numerous articles in professional journals. He has served as a U.N. NGO Commissioner for Disarmament and Peace Education, secretary of the Labor Studies section of the Latin American Studies Association, and Society for the Psychological Study of Social Issues (SPSSI) special expert at the U.N. Commission on Sustainable Development; and co-convened the seminar on Globalization and Class Relations at Colombia University. He has been past president and council delegate for AFT Local 2274, and a board member of the U.S. Labor Education in the Americas Project. He has received awards from the Organization of American States, the Fulbright Association, the Middle Atlantic Council of Latin American Studies, SPSSI, and the N.J. Peace Action. He has six children and nine grandchildren, and resides with his wife, Bette, in Montclair, New Jersey.